This series aims to report new developments in physical research and teaching – quickly, informally, and at a high level. The type of material considered for publication includes:

1. Preliminary drafts of original papers and monographs

2. Lectures on a new field, or presenting a new angle on a classical field

3. collections of seminar papers

4. Reports of meetings

Texts which are out of print but still in demand may also be considered if they fall within these categories.

The timeliness of a manuscript is more important than its form, which may be unfinished or tentative. Thus, in some instances, proofs may be merely outlined and results presented which have been or will later be published elsewhere.

Publication of *Lecture Notes* is intended as a service to the international physical community, in that a commercial publisher, Springer-Verlag, can offer a wider distribution to documents which would otherwise have a restricted readership. Once published and copyrighted, they can be documented in the scientific libraries.

Manuscripts

Manuscripts are reproduced by a photographic process; they must therefore be typed with extreme care. Symbols not on the typewriter should be inserted by hand in indelible black ink. Corrections to the typescript should be made by sticking the amended text over the old one, or by obliterating errors with white correcting fluid. The figures (in the original size) ready for reproduction should be inserted into the text. Should the text, or any part of it, have to be retyped, the author will be reimbursed upon publication of the volume. Authors receive 50 free copies.

The typescript is reduced slightly in size during reproduction, therefore a large size of type should be used; best results will not be obtained unless the text on any one page is kept within the overall limit of 18 x 26.5 cm (7 x 10½ inches). **The publishers will be pleased to supply on request special stationery with the typing area outlined.**

Manuscripts in English, German or French should be sent to Springer-Verlag, 6900 Heidelberg, Postfach 1780.

Die „*Lecture Notes*" sollen rasch und informell, aber auf hohem Niveau, über neue Entwicklungen in der Physik berichten. Zur Veröffentlichung kommen:

1. Vorläufige Fassungen von Originalarbeiten und Monographien.

2. Spezielle Vorlesungen über ein neues Gebiet oder ein klassisches Gebiet in neuer Betrachtungsweise.

3. Seminarausarbeitungen.

4. Vorträge von Tagungen.

Ferner kommen auch ältere vergriffene spezielle Vorlesungen, Seminare und Berichte in Frage, wenn nach ihnen eine anhaltende Nachfrage besteht.

Die Beiträge dürfen im Interesse einer größeren Aktualität durchaus den Charakter des Unfertigen und Vorläufigen haben. Sie brauchen Beweise unter Umständen nur zu skizzieren und dürfen auch Ergebnisse enthalten, die in ähnlicher Form schon erschienen sind oder später erscheinen sollen.

Die Herausgabe der „*Lecture Notes*" Serie durch den Springer-Verlag stellt eine Dienstleistung an die physikalischen Institute dar, indem der Springer-Verlag für ausreichende Lagerhaltung sorgt und einen großen internationalen Kreis von Interessenten erfassen kann. Durch Anzeigen in Fachzeitschriften, Aufnahme in Kataloge und durch Anmeldung zum Copyright sowie durch die Versendung von Besprechungsexemplaren wird eine lückenlose Dokumentation in den wissenschaftlichen Bibliotheken ermöglicht.

Lecture Notes
in Physics

Edited by J. Ehlers, München, K. Hepp, Zürich and
H. A. Weidenmüller, Heidelberg
Managing Editor: W. Beiglböck, Heidelberg

11

Othmar Steinmann
Schweizerisches Institut für Nuklearforschung, Zürich

Perturbation Expansions in Axiomatic Field Theory

Springer-Verlag Berlin Heidelberg GmbH 1971

ISBN 978-3-540-05698-0 ISBN 978-3-540-37023-9 (eBook)
DOI 10.1007/978-3-540-37023-9

 Library of Congress Catalog Card Number 72-183483.

Originally published by Springer-Verlag Berlin Heidelberg New York in 1971.

Offsetdruck: Julius Beltz, Hemsbach/Bergstr.

CONTENTS

I. INTRODUCTION

In view of the persistent lack of non-trivial exact models in relativistic quantum field theory, approximation schemes acquire an interest beyond their possible usefulness for numerical calculations. Their close study may reveal properties which are true also in an exact theory and may provide valuable clues for the search after such exact models.

Among the various methods of approximation hitherto developed, perturbation theory is certainly the one that has been investigated in most detail. It is distinguished by the fact that it can be refined to an arbitrarily high degree: it yields an infinite sequence of approximations of ever higher complexity. This increasing complexity represents, it is hoped, an ever closer approach to reality. Unfortunately this does not necessarily go together with an increasing numerical accuracy. The few indications we have tend to show that the sequence of successive perturbative approximations diverges, so that perturbation theory will presumably not lead to an existence proof for non-trivial models. Nevertheless, the study of perturbation theory is of interest for the reasons mentioned above - even apart from its numerical successes in electrodynamics and, to a lesser degree, in the weak interactions. In particular it can be used as a testing ground for conjectures about properties of field theoretical quantities. It is usually believed that properties which are not true in perturbation theory are also not true in an exact theory. (Needless to say, this criterion has to be used with a certain discretion.)

Traditionally perturbation theory is developed in the framework of canonical field theory, which originated in a simple generalisation of the rules of non-relativistic quantum mechanics to systems with an infinite number of degrees of freedom*. A specific form of

* For a thorough discussion of the problems and results of canonical perturbation theory we refer the reader to the Paris lectures of K.Hepp[1]. More formal accounts can be found in any of the well known textbooks on the subject.

the Hamiltonian or Lagrangian as function of the fields is postulated and the corresponding equations of motion are solved by expanding all relevant quantities into a power series with respect to a coupling constant. Unfortunately, many of the expansion coefficients turn out to diverge. They can be made finite by a procedure called "renormalisation" at the price of introducing compensation terms with infinite coefficients into the equations of motion and the Hamiltonian. In short: the original equations have no solutions, and the equations that do have solutions do not seem to make sense at first glance. The usual method of injecting some sense into them is that of defining the infinite coefficients as limits of finite quantities by first introducing various cut-offs and then removing them again. A more sophisticated method originally suggested by Valatin [2] has been highly developed in the last few years by Brandt, Wilson,and Zimmermann (see Refs.[3,4] and the original papers quoted there). They note that the divergences in the equations of motion are due to the occurrence of notoriously ill-defined products of fields at the same point and proceed to define these products as limits of operator products at different points. Both methods have the disadvantage that they rob the canonical formalism of its simplicity and intuitive appeal.

A different point of view has been taken by the Bogoliubov school [5]. They derive the rules of perturbation theory, including renormalisation, in a conceptually more satisfying way by refusing to take the canonical formalism too serious. A Lagrangian is used merely to determine the theory in first order of the coupling constant, where divergences do not yet appear. The higher orders are then calculated, not from equations of motion,but by making extensive use of the requirements of causality and unitarity of the S-matrix.

From here it is not far to the point of view that will be adopted in the present work: that of dispensing entirely with Lagrangians and canonical commutation relations for interacting fields and studying the problem in the framework of axiomatic field theory.

It has already been pointed out by Lehmann, Symanzik, and Zimmermann in the first of their papers establishing what has come to be known as the LSZ formalism [6], that their approach to field theory opens up a new way of introducing the dynamics, in particular of constructing perturbative expansions. They showed that a field theory can be completely characterised by its Green's functions (=time ordered functions), a set of distributions $\tau(x_1,..,x_n)$, n=2,3,... , satisfying certain conditions, in particular an infinite set of quadratic integral equations. They proposed to specify a particular model by prescribing some boundary conditions for the τ and showed how this idea worked in the lowest orders of perturbation theory. Later on they showed that the retarded functions $r(x_1,..,x_n)$ could be used instead of the τ [7,8].

These ideas have been taken up and developed in Refs. 9-11 . It was shown that in this way ultraviolet divergences can be avoided and are replaced by ambiguities in the solutions of the relevant equations. It was, however, not shown that the ultraviolet problem is not simply replaced by other problems of equal difficulty. Existence problems other than the UV question were ignored. In other words: a formalism was developed but not proved to work.

In the present work we purport to give such a proof, based on ideas put forward in an incomplete fashion in two earlier papers 12 . A perturbative formalism will be developed that allows one to calculate the retarded functions r in any order of perturbation theory by recursive solution of the unitarity equations amended with certain boundary conditions. It will be proved rigorously that the formalism yields finite results in all finite orders. No information is obtained on the vital question of the convergence of the perturbation series. Hence, in what follows, any statement of the type "$\underline{A}$ is true in perturbation theory" must be read as "$\underline{A}$ is true in every finite order of perturbation theory".

We assume that the reader is familiar with the basic ideas and the more important results of Wightman field theory as explained in the books by Streater and Wightman [13] and by Jost [14], as well as with the LSZ formalism [6-8]. For ease of reference the relevant definitions and results will be collected in Chapter II, but without motivations or proofs.

We consider only the simplest case, that of a single, scalar, hermitian field $A(x)$ associated with spinless particles of mass $m > 0$. The generalisation to any number of fields with arbitrary spin is straightforward. (My phenomenological friends inform me, however, that it is not trivial.) The restriction to massive particles is more serious. The LSZ formalism in its present form does not apply to massless particles, and a suitable generalisation is not in sight. This has the unfortunate consequence that quantum electrodynamics, the theory in which perturbation methods have been most successful, is not covered by our formalism.

As to the relation between the formalism to be explained here and canonical field theory we can say the following. The r - functions calculated in the canonical way satisfy all the requirements that we shall use to construct our r [1]*. Hence the canonical theories (for polynomial Lagrangians) are certainly contained in the set of models considered here. Given this fact it is then easy to find a 1-1-relation between the two categories of models (see Chapter IV): the two formalisms are equivalent. A comparison in detail of the two formulations has, however, not yet been carried out. Presumably such a comparison could be effected most easily with the help of yet another, in some sense intermediate, formulation of perturbation theory that has been developed recently by Epstein and Glaser [16,17]. These authors take over from canonical theory some formal expressions, the Gell-Mann-Low expansions, which give the retarded or time-ordered products respectively in terms of retarded or time-ordered products of certain Wick polynomials in free fields. They show that these products can be defined rigorously in such a way that the resulting r (or τ) satisfy all the axiomatic requirements. In other words: they solve the same problem as we do, but they do it by taking over the canonical expressions and showing that they have the correct properties, while we use the postulated properties to construct a solution from scratch.

The main advantage of our method over the canonical one is its logical simplicity and straightforwardness. We work from the start with the real problem in its full complexity. No temporary mutilation

---

* The one outstanding problem here, the existence of the adiabatic limit, has recently been solved by Hepp [15]. I thank K. Hepp for this information.

by cut-offs and no formal juggling with divergent quantities is required. In particular, we are from the very beginning in the adiabatic limit, so that the question of the existence of this limit never presents itself as an explicit problem. On the other hand the method is unsuited for numerical calculations. For such purposes the Feynman graphs of the traditional formulation are vastly superior. Of course, the two formalisms being equivalent, it should be possible to derive the Feynman rules in our framework. But as yet this has not been done.

Finally we wish to mention that a similar formalism could be developed using the time-ordered instead of the retarded products [9,17]. Some steps in our formalism could be simplified by making more extensive use of the generalised retarded products [16]. However, the ordinary retarded functions are more immediately related to the physical content of the theory since they are the coefficients in the Haag expansion of the interacting field (see Chapter II). We prefer therefore to base the formalism on them exclusively and use the generalised products only as auxiliary quantities in some of the proofs.

Remark on notation. The following notation will turn out to be useful. We shall often have to deal with functions of several variables of the same general character playing symmetrical roles in the theory. We introduce the convention that such related variables are denoted by the same small latin or greek character with distinguishing suffixes, e.g. $(u_1,\dots,u_n)$ or $(\xi_1,\dots,\xi_m)$, etc. All the variables in a set are of the same type: all scalars or all 4-vectors, etc. We represent such a set simply by the corresponding capital letter:

$$U = \{u_1,\dots,u_n\} \quad , \quad \Xi = \{\xi_1,\dots,\xi_m\} \tag{1.1}$$

and write the dependence of a function or distribution f on the set as

$$f(u_1,\dots,u_n) = f(U), \quad f(u_1,\dots,u_n,\xi_1,\dots,\xi_m) = f(U,\Xi) , \tag{1.2}$$

etc. Furthermore we define

$$\alpha U = \{\alpha u_1,\dots,\alpha u_n\} \tag{1.3}$$

for α any complex number,

$$|U|^2 = \sum_1^n |u_i|^2 , \qquad (1.4)$$

and

$$dU = \prod_1^n du_i . \qquad (1.5)$$

If u_i is a 4-vector, then $|u_i|$ is its Euclidean length, i.e. $|u_i|^2 = u_{io}^2 + |\underset{\sim}{u}_i|^2$. In the case of 3- or 4-vectors du_i is an abbreviation for $d^3\underset{\sim}{u}_i$ or d^4u_i respectively.

II. THE LSZ FORMALISM

In this chapter we shall give a compilation of the assumptions, definitions, and theorems of the LSZ formalism as far as we need them. This is meant to serve as a reminder and to fix the notations and conventions. Proofs and detailed explanations will in general not be given. We follow essentially Ref. [18], with some sign mistakes corrected.

We consider the theory of a scalar hermitian field $A(x)$ satisfying the Wightman axioms in the special form appropriate to the description of spinless particles with mass $m>0$. These axioms are the following.

Postulate 1 (quantum mechanics): The pure states of the theory are represented by vectors in a Hilbert space $\mathfrak{H}$, observables and other physical quantities by linear operators in $\mathfrak{H}$.

Postulate 2 (relativistic invariance): A strongly continuous unitary representation $U(\Lambda,a)$ of the connected Poincaré group $\mathcal{P}_+^\uparrow$ is defined on $\mathfrak{H}$. Λ is the homogeneous part, a the translation part of the Poincaré transformation (Λ,a).

Postulate 3 (spectrum): Let

$$P_\mu = \int p_\mu \, dE(p) \qquad (2.1)$$

be the infinitesimal generators of the representation $T(a) = U(\mathbf{1},a)$ of the translation group:

$$T(a) = \exp\{i\, P_\mu a^\mu\} . \qquad (2.2)$$

Then the support of the spectral measure $dE(p)$ consists of the isolated point $p = 0$, the 1-particle hyperboloid $p^2 = m^2$, $p_0>0$, and a continuum in $p^2>4m^2$, $p_0>0$. Let $\mathfrak{H}_0$, $\mathfrak{H}_1$ be the eigenspaces of the mass operator $M^2 = P_0{}^2 - \underset{\sim}{P}^2$ belonging to the eigenvalues 0 and m^2 respectively. $\mathfrak{H}_0$ is 1-dimensional. It is spanned by a normalised vector, the "vacuum", which will be denoted alternatively by Ω or $|0\rangle$. The restriction of $U(\Lambda,a)$ to $\mathfrak{H}_1$ forms an irreducible representation of the Poincaré group with mass m and spin 0.

<u>Postulate 4 (field theory)</u>: An operator valued tempered distribution with the following properties is defined in $\mathfrak{H}$.

a) To each tempered test function $\phi(x_1,\dots,x_n)\in\mathcal{S}$ [19], n arbitrary, corresponds a closable operator

$$A^n(\phi) = \int dX\, A(x_1)\dots A(x_n)\;\phi(X) \qquad (2.3)$$

called a <u>field monomial</u>. Note that $A^1(\phi) = A(\phi)$. All field monomials are defined on Ω, and the vector valued functional $V(\phi) = A^n(\phi)\,\Omega$ is continous in ϕ in the strong topology of $\mathfrak{H}$. The linear space D spanned by the vectors $A^n(\phi)\Omega$, $n = 0,1,2,\dots,$ is dense in $\mathfrak{H}$. Note that these assumptions imply that $A^n(\phi)$ is defined on D.

b) $A(\phi)$ is hermitian for real ϕ.

c) A transforms under $\mathcal{P}_+^\uparrow$ as a scalar:

$$A(\Lambda x+a) = U(\Lambda,a)\, A(x)\, U^*(\Lambda,a) . \qquad (2.4)$$

d) A is local:

$$[A(x),A(y)] = 0 \quad \text{if} \quad (x-y)^2 <0. \qquad (2.5)$$

(We use the metric in which time-like vectors have positive square.)

e) $A(x)$ has non-vanishing matrix elements between Ω and $\mathcal{H}_1$. A can then be normalised so that these matrix elements are equal to the corresponding ones for a free field. The vacuum expectation value of A shall vanish:

$$\langle 0|A(x)|0\rangle = 0. \tag{2.6}$$

This condition is no restriction of generality since it can always be achieved by addition of a constant c-number to $A(x)$.

Let

$$\tilde{A}(p) = (2\pi)^{-5/2}\int dx\, e^{ipx} A(x) \tag{2.7}$$

be the Fourier transform of A. Note that the (improper) vector $\tilde{A}(p)\Omega$ has momentum $-p$, while $\tilde{A}^*(p)\Omega$ has momentum p. This somewhat akward convention, as well as the strange (2π)-power in (2.7), is due to our desire to stick to the conventions most often used for free fields. The normalisation introduced in e) means that the following Källén-Lehmann representation for the 2-point function holds:

$$\langle 0|\tilde{A}(p)\,\tilde{A}(q)0\rangle = \delta^4(p+q)\;\{\delta_+(p) + \Theta(p_0)\sigma(p^2)\} \tag{2.8}$$

with

$$\delta_\pm(p) = \Theta(\pm p_0)\;\delta(p^2-m^2) \tag{2.9}$$

and σ a tempered measure with support in $p^2 \geq 4m^2$.

Under the stated assumptions the following asymptotic conditions can be proved.

Let $\mathcal{G}\subset\mathcal{S}$ be the space of the test functions $\tilde{f}(p)$ with support in the set

$$G = \{p : 0 \leq p^2 \leq 4m^2\;,\;\; p_0 \geq 0\}\,. \tag{2.10}$$

To $\tilde{f} \in \mathcal{G}$ we associate the operator valued function

$$A_f(t) = \int dp\; e^{-itp^-}\; \tilde{f}^*(p)\, \tilde{A}(p) \quad , \tag{2.11}$$

where

$$p^{\pm} = p_0 \pm \omega(\underset{\sim}{p}) \quad , \quad \omega(\underset{\sim}{p}) = +(\underset{\sim}{p}^2 + m^2)^{\frac{1}{2}} \quad . \tag{2.12}$$

Let $A^{ex}(x)$ be a scalar hermitian free field to mass m. $\tilde{A}^{ex}(p)$, defined as in (2.7), is of the form

$$\tilde{A}^{ex}(p) = \delta_+(p)\, \hat{A}^{ex}(\underset{\sim}{p}) + \delta_-(p)\, \hat{A}^{ex*}(-\underset{\sim}{p}) \quad , \tag{2.13}$$

with

$$\begin{aligned} [\hat{A}^{ex}(\underset{\sim}{p}), \hat{A}^{ex}(\underset{\sim}{q})] &= 0, \\ [\hat{A}^{ex}(\underset{\sim}{p}), \hat{A}^{ex*}(\underset{\sim}{q})] &= 2\omega(\underset{\sim}{p})\, \delta^3(\underset{\sim}{p}-\underset{\sim}{q}) \quad . \end{aligned} \tag{2.14}$$

To $\hat{f}(\underset{\sim}{p}) \in \mathcal{S}$ we associate the destruction operator

$$A_{\hat{f}}^{ex} = \int \frac{d^3p}{2\omega(\underset{\sim}{p})}\, \hat{f}^*(\underset{\sim}{p})\, \hat{A}^{ex}(\underset{\sim}{p}) \quad . \tag{2.15}$$

The asymptotic conditions state that two free fields A^{in} and A^{out} exist in $\mathcal{H}$ such that $A_f(t)$ converges for $t \to \pm\infty$ to $A_{\hat{f}}^{out}$ or $A_{\hat{f}}^{in}$ respectively, with

$$\hat{f}(\underset{\sim}{p}) = \tilde{f}(\omega(\underset{\sim}{p}), \underset{\sim}{p}) \quad , \tag{2.16}$$

in one of the following two senses.

<u>Haag-Ruelle asymptotic condition</u>. The vector

$$\Phi(t) = A_{f_1}^{(*)}(t) \ldots A_{f_n}^{(*)}(t)\Omega \tag{2.17}$$

converges strongly to

$$\Phi^{ex} = A_{\hat{f}_1}^{ex(*)} \ldots A_{\hat{f}_n}^{ex(*)}\, \Omega \quad : \tag{2.18}$$

$$\underset{t \to \pm\infty}{\text{st-lim}}\; \Phi(t) = \Phi^{ex} \quad , \tag{2.19}$$

where "ex" stands for "in" in the case $t\to-\infty$, and for "out" in the case $t\to\infty$. The symbol $^{(*)}$ means that an asterisk may or may not be present. Asterisks occur in the same places in (2.17) and (2.18).

LSZ asymptotic condition. A state Φ^{ex} of the form (2.18) is called "non-overlapping" if the supports of the wave functions $\hat{f}_i$ are mutually disjoint. Let $\mathcal{L}_0^{ex}$ be the linear hull of all non-overlapping ex-states. Then $A_f^{(*)}(t)$ is defined on $\mathcal{L}_0^{ex}$ and converges weakly to $A_f^{ex(*)}$ for $t\to\pm\infty$:

$$\underset{t\to\pm\infty}{\text{w-lim}}\, A_f^{(*)}(t)\, \Phi^{ex} = A_{\hat{f}}^{ex^{(*)}}\, \Phi^{ex} \qquad (2.20)$$

for $\Phi^{ex}\in\mathcal{L}_0^{ex}$. Again "ex" stands for "in" or "out" as the case may be. The restriction to non-overlapping states can be dropped in perturbation theory. There $A_f^{(*)}(t)$ exists on all states of the form (2.18) irrespective of the supports of the $\hat{f}_i$ and converges as in (2.20). Only this situation will be considered further.

Let $\mathcal{H}^{ex}$ be the Fock space of A^{ex}, i.e. the Hilbert space spanned by the Φ^{ex} of (2.18) with $n = 0,1,2,\ldots$ We demand

Postulate 5 (asymptotic completeness):

$$\mathcal{H}^{in} = \mathcal{H}^{out} = \mathcal{H} \quad . \qquad (2.21)$$

Important tools of the LSZ formalism are the retarded products $R(x_1,..,x_n)$ which are defined by the following conditions.

i) $R(x_1,..,x_n)$ is an operator-valued tempered distribution. Its value $R(\phi)$ on a test function $\phi(x_1,..,x_n)\in\mathcal{S}$ has the properties of a field monomial as stated in Postulate 4a, i.e. it is a closable operator which is defined on D and maps D into itself.

ii) $R(\phi)$ is hermitian for real ϕ .

iii) $$R(x_1) = A(x_1).$$

iv) $R(x_1,x_2,..,x_n)$ is invariant under permutations of the arguments $x_2,\ldots,x_n$.

v) The R satisfy the identities

$$R(x,y,x_1,..x_n) - R(y,x,x_1,..,x_n) =$$

$$= -i \ \Sigma \ \left[R(x,X_L),R(y,X_R)\right] \quad , \qquad (2.22)$$

where the summation extends over all partitions of the set $X=\{x_1,..,x_n\}$ into two complementary subsets X_L and X_R , one of which may be empty.

vi) The support of $R(x_1,..,x_n)$ is contained in the set

$$T_n = \{(x_1,..,x_n) : (x_1-x_i) \in \bar{V}_+ \text{ for } i=2,..,n\} \quad , \qquad (2.23)$$

$\bar{V}_+$ the closed forward cone.

vii) The R are covariant:

$$R(\Lambda x_1+a,...,\Lambda x_n+a) = U(\Lambda,a)\ R(x_1,..,x_n)\ U^*(\Lambda,a) \ . \qquad (2.24)$$

Objects satisfying these conditions are called "sharp" retarded products. It is not known at present whether they always exist in a Wightman theory satisfying Postulates 1)-5). As yet it has only been possible in this general framework to prove the existence of "smooth" retarded products. They are distinguished from the sharp ones in that their support lies in an ε-neighbourhood of T_n and that the covariance equation (2.24) holds only for translations, not for Lorentz transformations. In perturbation theory, however, sharp R do exist, so that we need not worry about this point.

Conditions i)-vii) do not define the R uniquely. This is unimportant. Any set of R satisfying i)-vii) will do.

The vacuum expectation value of R is called a <u>retarded function</u> and is denoted by r :

$$r(X) = \langle 0|R(X)|0\rangle \ . \qquad (2.25)$$

Define the Fourier transform $\tilde{r}$ of r by

$$\tilde{r}(p_1,..,p_n) = (2\pi)^{-\frac{5}{2}n} \int dX \ \exp\{i\Sigma p_j x_j\} r(x_1,..,x_n). \qquad (2.26)$$

The Fourier transform $\tilde{R}$ of R is defined analogously. $\tilde{r}$ is of the form

$$\tilde{r}(p_1,..,p_n) = \delta^4(\Sigma p_j)\ \hat{r}(p_1,..,p_n) \quad . \tag{2.27}$$

The δ-factor expresses energy-momentum conservation. The function $\hat{r}$ is only defined on the manifold $\Sigma p_j = 0$, so that one of the variables is redundant.

"Amputation" of $\tilde{r}$ with respect to the variable p_i means multiplication of $\tilde{r}$ with $(p_i^{\,2} - m^2)$. In x-space this corresponds to application of a Klein-Gordon operator $-(\Box_{x_i} + m^2)$. Define

$$\tilde{r}^{amp}(p_1,..,p_n;q_1,..,q_\ell) = \prod_{i=1}^{\ell} (q_i^{\,2} - m^2)\tilde{r}(p_1,..,q_\ell) \ , \tag{2.28}$$

so that the variables standing to the right of the semicolon are amputated.

Consider $\tilde{r}^{amp}(p_1,..,p_n;q_1,..,q_\alpha,q_1',..,q_\beta')$ in a neighbourhood of $q_i^- = 0$, $q_i'^+ = 0$, i.e. for q_i near the positive mass shell, q_i' near the negative mass shell. We can introduce q_i^-, $q_i'^+$ as new variables instead of q_{i_0}, q_{i_0}' . The resulting distribution will still be called $\tilde{r}^{amp}$ by abuse of notation. Let $\phi(P), f(\underset{\sim}{Q}), g(\underset{\sim}{Q}')$ be tempered test functions and define

$$\rho(Q^-,Q'^+) = \int dP\, d\underset{\sim}{Q}\, d\underset{\sim}{Q}'\ \phi(P)\ f(\underset{\sim}{Q})\ g(\underset{\sim}{Q}')\tilde{r}^{amp}(P;Q^-,\underset{\sim}{Q},Q'^+,\underset{\sim}{Q}') \,. \tag{2.29}$$

We assume that ρ is a continous function in a neighbourhood of the origin. This is true in perturbation theory. In general it has only been proved to be a consequence of Postulates 1-4 if the product fg vanishes with all derivatives in the points where two $\underset{\sim}{q}_i$ or two $\underset{\sim}{q}_i'$ or a $\underset{\sim}{q}_i$ and a $-\underset{\sim}{q}_j'$ coincide. [20] .

Let

$$\Phi_{in} = A_{\hat{f}_1}^{in*} \dots A_{\hat{f}_n}^{in*}\,\Omega \quad , \qquad \Psi_{in} = A_{\hat{g}_1}^{in*} \dots A_{\hat{g}_m}^{in*}\,\Omega \tag{2.30}$$

be two states of the form (2.18). The matrix element

$$M = (\,\Phi_{in}\ ,\ \tilde{R}(p_1,..,p_\ell)\ \Psi_{in}) \tag{2.31}$$

can be expressed in terms of the $\tilde{r}^{amp}$ with the help of the reduction formula

$$M = \sum_{\nu=0}^{\min(n,m)} \sum_{(\alpha_i,\beta_i)} \prod_{i=1}^{\nu} (\hat{f}_{\alpha_i},\hat{g}_{\beta_i})(\Phi_{in}(\alpha_i),\tilde{R}(P)\Psi_{in}(\beta_i))^T . \quad (2.32)$$

Here the second sum extends over all possible ways of pairing ν indices α_i , $1\leq\alpha_i\leq n$, with ν indices β_i , $1\leq\beta_i\leq m$. $\Phi_{in}(\alpha_i)$ and $\Psi_{in}(\beta_i)$ are obtained from Φ_{in} and Ψ_{in} respectively by omitting the creation operators whose wave functions $\hat{f}_j$, $\hat{g}_h$ occur in one of the factors $(\hat{f}_{\alpha_i} , \hat{g}_{\beta_i})$. These factors are

$$(\hat{f},\hat{g}) = \int \frac{d^3p}{2\omega(\underset{\sim}{p})} \hat{f}^*(\underset{\sim}{p})\,\hat{g}(\underset{\sim}{p}) . \quad (2.33)$$

Finally,

$$(\Phi_{in}, \tilde{R}(P)\,\Psi_{in})^T$$

$$= (2\pi)^{n+m} \int \prod_{j=1}^{n} [dq_j\ \delta_+(q_j)\ \hat{f}^*_j(\underset{\sim}{q}_j)] \prod_{h=1}^{m} [dq'_h\ \delta_+(q'_h)\ \hat{g}_h(\underset{\sim}{q}'_h)]\ \tilde{r}^{amp}(P;Q,-Q'). \quad (2.34)$$

According to assumption (2.29) this integral exists as a distribution in P .

From the identities (2.22) we obtain by taking their vacuum expectation value, summing on the right over a complete set of intermediate in-states, and using the reduction formula, the completeness equations (or GLZ equations) [8]

$$\tilde{r}(p,q,P)-\tilde{r}(q,p,P) = -i \sum_{L,R} \sum_{\ell=1}^{\infty} \frac{(2\pi)^{2\ell}}{\ell!} \int \prod_1^{\ell} dk_i \{ \prod_1^{\ell}\delta_+(k_i) - \prod_1^{\ell} \delta_-(k_i) \}$$

$$\times\ \tilde{r}^{amp}(p,P_L;-K)\ \tilde{r}^{amp}(q,P_R;K) . \quad (2.35)$$

Here P stands for the set $\{p_1,..,p_n\}$ and the sum $\Sigma_{L,R}$ extends over all partitions of P into two complementary subsets P_L and P_R .

In x-space these equations become

$$r(x,y,X) - r(y,x,X) = -i \sum_{L,R} \sum_{\ell=1}^{\infty} \frac{i^{\ell}}{\ell!} \int \prod_{1}^{\ell} [du_i\ dv_i]\ K^{\ell}(U-V)$$

$$\times\ r^{amp}(x,X_L;U)\ r^{amp}(y,X_R;V) \tag{2.36}$$

with $X = \{x_1,..,x_n\}$, $U = \{u_1,..,u_\ell\}$ etc. and

$$K^{\ell}(U-V) = \prod_{1}^{\ell} \Delta_+(u_i-v_i) - \prod_{1}^{\ell} \Delta_+(v_i-u_i) \ , \tag{2.37}$$

$$\Delta_+(\xi) = -\frac{i}{(2\pi)^3}\int d^4p\ \delta_+(p)\ e^{-ip\xi} \ . \tag{2.38}$$

The central element of our approach to perturbation theory is the GLZ theorem [8,18]. It states that a field theory as described above can be fully characterised by its retarded functions and gives the conditions under which a given set of retarded functions defines a field theory. We shall give a version of the theorem which is specially adapted to the needs of perturbation theory. The assumptions as formulated here are stronger than necessary. But they turn out to be satisfied in perturbation theory and have the advantage of simplifying the formulation of the theorem considerably.

Before we can formulate the theorem we have to introduce some new concepts and notations.

a) A function $f(u_1,..,u_n)$ is called Hölder continuous of index ε, $\varepsilon > 0$, if

$$\frac{f(u_1+a_1,..,u_n+a_n) - f(u_1,..,u_n)}{|A|^{\varepsilon}} \qquad |A|^2 = \Sigma\ a_i^{\ 2},$$

remains bounded for $|A| \to 0$ for all u_i. We introduce the space H_ε of Hölder continuous functions with strong decrease at infinity. Its elements are the functions $f(U)$ for which

$$\|f\|_N = \sup_{U,|A|\leq A_0} \{(1+U^2)^N \left[|f(U)| + \frac{|f(U+A) - f(U)|}{|A|^{\varepsilon}}\right\} < \infty \tag{2.39}$$

for an arbitrary finite A_0 and all positive integers N.*

b) In (2.15) we assumed that the wave function $\hat{f}$ is a tempered test function. In fact the operators $A^{ex}_{\hat{f}}$, A^{ex} a free field, exist for a much larger class of wave functions, namely for all $\hat{f}$ with

$$\int \frac{d^3p}{2\omega(\underline{p})} |\hat{f}(\underline{p})|^2 < \infty .$$

This space contains in particular the functions $\hat{f}(\underline{p}) \in H_\varepsilon$, $\varepsilon > 0$. Moreover, for $\hat{f}(\underline{p}_1,..,\underline{p}_n) \in H_\varepsilon$, the vector

$$\Phi^{ex}_{\hat{f}} = \int \frac{d^3p_1}{2\omega(\underline{p}_1)} \cdots \frac{d^3p_n}{2\omega(\underline{p}_n)} \hat{f}(\underline{p}_1,..,\underline{p}_n) \hat{A}^{ex*}(\underline{p}_1) \ldots \hat{A}^{ex*}(\underline{p}_n)\, \Omega \tag{2.40}$$

exists. We call $\mathcal{L}^{ex}_\varepsilon$ the linear space spanned by all vectors of this form.

c) We introduce an auxiliary function $\chi(p)$ of the 4-vector p with the following properties : χ is C^∞ , $\chi = 1$ for $p_0 = \omega(\underline{p})$, and

$$\mathrm{supp}\chi \subset \{p : p_0 > 0 , \quad 0 < A \leq p^2 \leq B < 4m^2\} , \tag{2.41}$$

where A,B are chosen such that $(p+q) \notin \mathrm{supp}\chi$ if $p,q \in \mathrm{supp}\chi$. χ can be chosen to be of the form

$$\chi(p) = \theta(p_0)\bar{\chi}(p^2) \tag{2.42}$$

with $\bar{\chi}(m^2) = 1$ and $\mathrm{supp}\bar{\chi} \subset \{A \leq p^2 \leq B\}$.

We can now formulate our version of the GLZ theorem.

---

* Hölder continuous wave functions have been introduced by Schneider [21] . For the properties of Hölder continuous functions we refer to this paper and to [22,23] .

Theorem 2.1

Let $r(x_1,..,x_n)$, $n=2,3,...$, be tempered distributions with the following properties

i) $r(X)$ is real, invariant under $\mathcal{P}_+^\uparrow$, and invariant under permutations of the variables $x_2,..,x_n$.

ii) The support of $r(X)$ is contained in T_n as defined in (2.23).

iii) Let $0<\varepsilon<\frac{1}{2}$. Let $f(p_1^-,\underset{\sim}{p}_1,...,p_\beta^-,\underset{\sim}{p}_\beta,q_1,...,q_\gamma)\in H_\varepsilon$, $\phi(k_1,..,k_\alpha)\in\mathcal{S}$

Then

$$F(P^-,Q) = \prod_1^\gamma \chi(-q_h) \int dK\, d\underset{\sim}{P} \prod_1^\beta \chi(p_j)\, \phi(K)\, f(P^-,\underset{\sim}{P},Q)$$

$$\times\ \mathcal{R}^{amp}(K;P^-,\underset{\sim}{P},Q) \tag{2.43}$$

exists and is in $H_{\varepsilon'}$ for any $\varepsilon'<\varepsilon$. Note that all the functions in the integrand must be written as functions of P^-, $\underset{\sim}{P}$ rather than P_o, $\underset{\sim}{P}$. The case $\alpha = 0$, $\beta+\gamma=2$ is excluded.

iv) The r satisfy the completeness equations (2.36).

Under these assumptions the Haag series

$$A(x) = \sum_{\ell=1}^{\infty} \frac{1}{\ell!} \int du_1..du_\ell\, r^{amp}(x;u_1,..,u_\ell)\, :A^{in}(u_1)..A^{in}(u_\ell): \tag{2.44}$$

converges strongly on $\mathcal{L}_\varepsilon^{in}$ and defines a Wightman field satisfying Postulates 1-5 whose retarded functions are the given ones. More exactly : the given r are a possible set of retarded functions for A. (This qualification is necessary because of the non-uniqueness of the definition of r.)

Condition iii) guarantees the existence of the integrals occurring in (2.36). Condition iv) contains in particular the assumption that the sum in (2.36) converges. More exactly we assume that the ℓ-sum converges separately for each partition (L,R). The double dots : : in (2.44) denote Wick ordering.

The expansion (2.44) can be generalised to

$$R(x_1,..,x_n) = r(x_1,..,x_n) + \sum_{\ell=1}^{\infty} \frac{1}{\ell!} \int du_1..du_\ell$$

$$\times\ r^{amp}(x_1,..,x_n;u_1,..,u_\ell)\, :A^{in}(u_1)..A^{in}(u_\ell): \tag{2.45}$$

The proof of this form of the GLZ theorem can be easily adapted from the proof of the more general version given in [18]. Our strong form of assumption iii) makes it unnecessary to introduce the generalised retarded functions in the formulation of the theorem, even though they are still very useful for the proof. It can be shown that they exist under our assumptions and have the correct properties.

Since the combinatorial part of the proof has been given rather short shrift in [18], as well as in the original paper [8], we shall supply here its main portion: we prove that the retarded operators (2.45) satisfy the identities (2.22). This proof serves also as a model for similar calculations that will be needed later on but will not be given explicitly.

For simplicity we consider the 2-point case

$$R(x,y) - R(y,x) = -i\,[A(x),A(y)] \quad . \tag{2.46}$$

The extension to the general case is straightforward.

We calculate the r.h.s. of (2.46) by inserting the expansion (2.44):

$$\begin{aligned} \text{r.h.s.} &= -i\,[A(x),A(y)] \\ &= -i\{\sum_{\ell=0}^{\infty}\sum_{\ell'=0}^{\infty}\frac{1}{\ell!}\frac{1}{\ell'!}\int\prod_{1}^{\ell}du_i\prod_{1}^{\ell'}dv_i\; r^{amp}(x;U)\; r^{amp}(y;V) \\ &\times\; :A^{in}(u_1)..A^{in}(u_\ell):\; :A^{in}(v_1)..A^{in}(v_{\ell'}): - (x\leftrightarrow y)\} \quad . \end{aligned} \tag{2.47}$$

According to Wick's theorem we have

$$\begin{aligned} :A^{in}(u_1)..A^{in}(u_\ell):\; :A^{in}(v_1)..A^{in}(v_{\ell'}): &= \sum_{k=0}^{\min(\ell,\ell')} i^k \\ \times \sum_{(i_\nu,j_\nu)} \Delta_+(u_{i_1}-v_{j_1})\,..\,\Delta_+(u_{i_k}-v_{j_k})\; &\overline{:A^{in}(u_1)..A^{in}(v_{\ell'}):} \quad . \end{aligned} \tag{2.48}$$

The second sum extends over all possible pairings of k u_i with k v_j. The bar over : : denotes omission of the A^{in} whose arguments occur in one of the Δ_+-factors. We note that in view of the symmetry of r

all the terms with a fixed k in (2.48) give the same contribution to (2.47). It is therefore sufficient to calculate the contribution of one term only and multiply it with the number $\binom{\ell}{k}\binom{\ell'}{k} k!$ of terms:

$$\text{r.h.s.} = -i \{ \sum_{\ell=0}^{\infty} \sum_{\ell'=0}^{\infty} \sum_{k=0}^{\min(\ell,\ell')} i^k \frac{1}{k!} \frac{1}{(\ell-k)!} \frac{1}{(\ell'-k)!} \int \prod_1^{\ell} du_i \prod_1^{\ell'} dv_j$$

$$\times \prod_{\nu=1}^{k} \Delta_+(u_\nu - v_\nu)\, r^{amp}(x;U)\, r^{amp}(y;V) :A^{in}(u_{k+1})\dots A^{in}(u_\ell)$$

$$\times A^{in}(v_{k+1})\dots A^{in}(v_{\ell'}): - (x \leftrightarrow y) \} .$$

We rename the variables: $(u_{k+1},\dots,u_\ell,v_{k+1},\dots,v_{\ell'}) \to (w_1,\dots,w_a,w_{a+1},\dots,w_{a+b})$, $a=\ell-k$, $b=\ell'-k$, and interchange the order of summations in a way that can be shown to be permitted in the context in which this calculation is used. This gives

$$\text{r.h.s.} = -i \{ \sum_{k=0}^{\infty} \sum_{a=0}^{\infty} \sum_{b=0}^{\infty} i^k \frac{1}{k!} \frac{1}{a!} \frac{1}{b!} \int \prod_1^{a+b} dw_i :A^{in}(w_1)\dots A^{in}(w_{a+b}):$$

$$\times \prod_1^{k} \left[du_i\, dv_i\, \Delta_+(u_i - v_i)\right] r^{amp}(x;w_1,\dots,w_a,u_1,\dots,u_k)$$

$$\times r^{amp}(y;w_{a+1},\dots,w_{a+b},v_1,\dots,v_k) - (x \leftrightarrow y) \}$$

$$= -i \{ \sum_{\alpha=0}^{\infty} \frac{1}{\alpha!} \sum_{a=0}^{\alpha} \binom{\alpha}{a} \int \prod_1^{\alpha} dw_i :A^{in}(w_1)\dots A^{in}(w_\alpha):$$

$$\times \sum_{k=0}^{\infty} \frac{i^k}{k!} \int \prod_1^{k} \left[du_i\, dv_i\, \Delta_+(u_i - v_i)\right] r^{amp}(x;w_1,\dots,w_a,u_1,\dots,u_k)$$

$$\times r^{amp}(y;w_{a+1},\dots,w_\alpha,v_1,\dots,w_k) - (x \leftrightarrow y) \} .$$

Because of the symmetry of the Wick product we can replace $\Sigma_{a=0}^{\alpha} \binom{\alpha}{a}$ by $\Sigma_{L,R}$, the sum over all partitions of $W = \{w_1,\dots,w_\alpha\}$ into two complementary subsets W_L and W_R :

$$\text{r.h.s.} = -i \sum_{\alpha=0}^{\infty} \frac{1}{\alpha!} \int dW \, :A^{in}(w_1)\ldots A^{in}(w_\alpha): \{ \sum_{k=0}^{\infty} \frac{i^k}{k!} \sum_{L,R}$$

$$\times \int \prod_1^k \left[du_i \, dv_i \, \Delta_+(u_i - v_i)\right] r^{amp}(x;W_L,U) \, r^{amp}(y;W_R,V) - (x \leftrightarrow y) \} \; ,$$

and this becomes with (2.36)

$$\text{r.h.s.}$$

$$= \sum_{\alpha=0}^{\infty} \frac{1}{\alpha!} \int \prod_1^{\alpha} dw_i \; :A^{in}(w_1)\ldots A^{in}(w_\alpha): \{r^{amp}(x,y;W) - r^{amp}(y,x;W)\}$$

$$= R(x,y) - R(y,x) \quad ,$$

q.e.d.

The GLZ theorem as stated in Theorem 2.1 is not yet in the form in which we shall use it, but has to be reformulated in terms of the totally amputated r-functions. We define

$$r^{Amp}(x_1,..,x_n) = (-1)^n \prod_1^n (\Box_{x_i} + m^2) \, r(x_1,..,x_n) \quad ,$$

$$\tilde{r}^{Amp}(p_1,..,p_n) = \prod_1^n (p_i{}^2 - m^2) \, \tilde{r}(p_1,..,p_n) \quad . \tag{2.49}$$

The r^{Amp} are tempered distributions with the same invariance and support properties as r . The original r can be recovered from r^{Amp} by

$$r(x_1,\ldots,x_n)$$

$$= \int dy_1..dy_n \, \Delta_R(x_1-y_1) \, \Delta_R(y_2-x_2)\ldots\Delta_R(y_n-x_n) \; r^{Amp}(y_1,..,y_n) \quad ,$$

$$\tilde{r}(p_1,..,p_n) = (p_1{}^2-m^2+i\varepsilon p_{1o})^{-1} \prod_{i=2}^{n} (p_i{}^2-m^2-i\varepsilon p_{io})^{-1} \, \tilde{r}^{Amp}(p_1,..,p_n)$$

(2.50)

with

$$\Delta_R(\xi) = (2\pi)^{-4} \int dp \frac{e^{-ip\xi}}{p^2-m^2+i\varepsilon p_o} \quad . \tag{2.51}$$

The convolution in the x-space form of (2.50) exists and has the correct support and symmetries if this is the case for r^{Amp} .

Completeness equations for r^{Amp} can be obtained from (2.36) by applying Klein-Gordon operators in x, y, $x_1,\ldots,x_n$ on both sides. This results simply in a replacement of r and r^{amp} by r^{Amp} and of the semicolons in r^{amp} by commas. The same is, of course, true for the p-space form (2.35). An essential feature of these amputated GLZ equations is that the 2-point function has disappeared from the right-hand side. To see this, consider the case $n=0$ of (2.35) in the region $p^2 < 4m^2$:

$$\tilde{r}(p,q) - \tilde{r}(q,p) =$$

$$= -i\ (2\pi)^2 \int dk\ \{\delta_+(k) - \delta_-(k)\}\ \tilde{r}^{amp}(p;-k)\ \tilde{r}^{amp}(q;k) \quad . \tag{2.52}$$

The $\ell > 1$ terms do not contribute in this region because of the support $q+\Sigma k_i = 0$ of $\tilde{r}^{amp}(q;k_1,..,k_\ell)$ and the support of δ_+ . We use (2.27) and define

$$\bar{r}(q) = \hat{r}(-q,q) \quad . \tag{2.53}$$

(2.52) becomes

$$\bar{r}(q) - \bar{r}(-q) = -i\ (2\pi)^2\ \{\delta_-(q) - \delta_+(q)\}\ \bar{r}^{amp}(q)\ \bar{r}^{amp}(-q) \quad , \tag{2.54}$$

with

$$\bar{r}^{amp}(q) = (q^2-m^2)\ \bar{r}(q) \quad . \tag{2.55}$$

$\bar{r}(q)$ is the Fourier transform of $r(\xi,0)$:

$$\tilde{r}(q) = (2\pi)^{-1} \int d\xi \quad e^{-iq\xi} \quad r(\xi,0) \quad . \qquad (2.56)$$

Due to the support of $r(\xi,0)$, $\tilde{r}(q)$ is the boundary value of a function analytic in $\mathrm{Im}\, q \in V_-$. From this and (2.54) it follows with familiar techniques (edge-of-the-wedge theorem, see Ref. [13] , p. 74) that we must have

$$\tilde{r}(q) = \frac{1}{2\pi} \frac{1}{q^2-m^2-i\varepsilon q_o} + F(q) \quad , \qquad (2.57)$$

with $F(q)$ analytic in $q^2 < 4m^2$. Hence

$$\tilde{r}^{Amp}(p,q) = \delta^4(p+q) \{ \frac{1}{2\pi} (q^2-m^2) + (q^2-m^2)^2 F(q) \} . \qquad (2.58)$$

Take now the amputated form of (2.35) and consider the terms on the right which contain 2-point functions. They are the $\ell=1$ terms with $P_R = \emptyset$ or $P_L = \emptyset$, with $\emptyset$ the empty set. Their integrand contains a factor $\delta_\pm(k)\, \delta^4(p-k)\, (p^2-m^2)$ or $\delta_\pm(k)\, \delta^4(q+k)\, (q^2-m^2)$ respectively, and this vanishes.

As a side remark we note that as a consequence of (2.27) and (2.57) the $\ell=1$ term in the expansion (2.44) is simply $A^{in}(x)$.

We can now formulate the amputated version of the GLZ theorem.

Theorem 2.2

Let $r^{Amp}(x_1,\ldots,x_n)$ be tempered distributions with the properties i), ii), and iii) of Theorem 2.1 , which satisfy the amputated GLZ equations. The 2-point function $\tilde{r}^{Amp}(p,q)$ shall be of the form (2.58).

Then the $r(x_1,\ldots,x_n)$ defined by (2.50) satisfy the assumptions of Theorem 2.1 .

Proof: We have already noted that the integrals (2.50) exist. It is trivial to see that conditions i), ii), and iii) of Theorem 2.1 are satisfied. The only non-trivial problem is that of showing that the amputated GLZ equations together with the normalisation condition (2.58) imply the original GLZ equations. In order to see this we divide the amputated form of (2.35) by $(p^2-m^2+i\varepsilon p_o)(q^2-m^2+i\varepsilon q_o)\Pi_{j=1}^{n}(p_j{}^2-m^2-i\varepsilon p_{jo})$.

All the terms on the right go over into the corresponding unamputated terms, because of (2.50) . We obtain in this way all the terms on the right in (2.35) except the terms containing 2-point functions. Consider the term $\tilde{r}^{Amp}(p,q,P)$ on the left. It goes over into

$$\tilde{r}^{Amp}(p,q,P)\ \{(p^2-m^2+i\varepsilon p_o)\ (q^2-m^2+i\varepsilon q_o)\ \prod_1^n\ (p_j{}^2-m^2-i\varepsilon p_{jo})\}^{-1}$$

$$= \tilde{r}(p,q,P)\ -\ 2\pi i\ \{\ \delta_+(q)\ -\ \delta_-(q)\}\ \tilde{r}^{amp}(p,P;q)\quad .$$

The first term is what we need on the left, the second term can be moved to the right-hand side and becomes with (2.58)

$$-2\pi i \int dk\ \ \{\delta_+(k)\ -\ \delta_-(k)\}\ \tilde{r}^{amp}(p,P;-k)\ \tilde{r}^{amp}(q;k)\quad .$$

This is exactly the missing term with the second factor a 2-point function. In the same way the $\tilde{r}^{Amp}(q,p,P)$ term supplies $\tilde{r}(q,p,P)$ on the left and the term with the first factor a 2-point function on the right.

From now on we shall work exclusively with r^{Amp} . We can therefore drop the suffix Amp , with the understanding that r and $\tilde{r}$ will henceforth denote the totally amputated retarded functions. R and $\tilde{R}$ will from now on likewise be the totally amputated retarded products, in particular

$$R(x) = -\ (\square + m^2)\ A(x)\quad . \qquad (2.59)$$

Besides the ordinary retarded products introduced above we shall also use, in an auxiliary role, the <u>generalised retarded products</u> (g.r.p. for short) which will now be defined and their relevant properties discussed [18, 24-27] .

Let S be the index set $\{1,\ldots,n\}$. A <u>n-cell</u> in S is a sign-valued function σ_I = + or - , defined on the proper subsets I of S , such that (with $\complement I = S-I$)

$$\sigma_I \neq \sigma_{\complement I} \quad ,$$

$$\sigma_{I' \cup I''} = \sigma_{I'} \quad \text{if} \quad I' \cap I'' = \emptyset \text{ and } \sigma_{I'} = \sigma_{I''} \quad . \qquad (2.60)$$

Two n-cells are called <u>adjacent</u> if all σ_I except two coincide, in the two cells. The differing signs belong to complementary subsets I_o , $\complement I_o = S-I_o$, called the "borders" between the two cells.

With each n-cell C_μ we associate a g.r.p. $G_\mu(x_1,..,x_n)$ such that the following two conditions are satisfied.

a) Let C_μ , C_ν be adjacent, with border $I_o = \{i_1,...,i_k\}$. Let σ_{I_o} be negative in C_μ , positive in C_ν . We define a k-cell C_α in I_o by attributing to each proper subset of I_o the sign that the same subset has in C_μ . A(n-k)-cell C_β is defined in $\complement I_o$ in the same way. Then we demand

$$G_\mu(X) - G_\nu(X) = -i \left[G_\alpha(x_{i_1},..,x_{i_k}), G_\beta(x_{i_{k+1}},..,x_{i_n}) \right] \quad . \qquad (2.61)$$

b) A particular n-cell is given by $\sigma_i = +$ for $i=2,...,n$, where σ_i is the sign attached to the one-element set $\{x_i\}$. The corresponding g.r.p. shall be the ordinary retarded product $R(X)$.

These two conditions define the G_μ uniquely. They are covariant, hermitian, operator-valued distributions, i.e. they satisfy conditions i), ii), and vii) of the definition of R . Remember that R is totally amputated, hence the G_μ are also totally amputated.

Any two n-cells C_μ , C_ν can be connected through a chain C_μ , C_{μ_1} ,....., C_{μ_r} , C_ν in which neighbours are always adjacent. Equation (2.61) can then be generalised to

$$G_\mu(X) - G_\nu(X) = -i \sum_{\alpha,\beta} c^{\mu\nu}_{\alpha\beta} \left[G_\alpha(X_\alpha), G_\beta(X_\beta) \right] \quad . \qquad (2.62)$$

The sum $\Sigma_{\alpha,\beta}$ goes over a finite number of terms, X_α and X_β are complementary subsets of $X = \{x_1,...x_n\}$, and the $c^{\mu\nu}_{\alpha\beta}$ are real numbers. Equation (2.62) gives the difference of two g.r.p. of the same order in terms of commutators of g.r.p. of lower order. This representation is not unique, since several chains $C_\mu, \ldots, C_\nu$ of the required type exist in general.

In particular, we can choose G_ν to be the ordinary product $R(X)$:

$$G_\mu(X) = R(X) - i \sum_{\alpha,\beta} c^\mu_{\alpha\beta} \left[G_\alpha(X_\alpha), G_\beta(X_\beta)\right] \quad . \tag{2.63}$$

By iteration we obtain from this a representation of G_μ in terms of multiple commutators of ordinary retarded products:

$$G_\mu(X) = R(X) + \sum i^{\ell-1} c^\mu_{\alpha_1\ldots\alpha_\ell} \left[\ldots\left[R(X_{\alpha_1}), R(X_{\alpha_2})\right], \ldots, R(X_{\alpha_\ell})\right] . \tag{2.64}$$

The sum extends over the partitions of X into $2 \leq \ell \leq n$ non-overlapping subsets. The $c^\mu_{\alpha_1\ldots\alpha_\ell}$ are real. Some of the $c^\mu_{\ldots}$ may, of course, be zero.

Let C_μ be an n-cell. We define two (n+1)-cells $C^\pm_\mu$ by these rules: the σ_I for proper subsets I of $\{1,\ldots,n\}$ shall be the same as in C_μ , and σ_{n+1} is positive in C^+_μ , negative in C^-_μ . The corresponding g.r.p. are called $G^\pm_\mu(x_1,\ldots,x_n;x_{n+1})$. Iteration of this procedure gives

$$G^\pm_\mu(x_1,\ldots,x_n;x_{n+1},\ldots,x_{n+m}) = G^{\pm\ldots\pm}_\mu(x_1,\ldots x_n;x_{n+1};\ldots;x_{n+m}) \quad , \tag{2.65}$$

where all the signs have to be equal (all positive or all negative). $G^\pm_\mu(\ldots;\ldots)$ is invariant under permutations of the variables standing to the right of the semicolon. For the ordinary retarded products we have

$$R^+(X;Y) = R(X,Y) \quad . \tag{2.66}$$

The vacuum expectation value $g_\mu(X)$ of $G_\mu(X)$ is called a <u>generalised</u> <u>retarded</u> <u>function</u> (g.r.f.) .

The g.r.p. have similar reduction formulae as R . The matrix element

$$M = (\Phi_{in}, \tilde{G}_\mu(p_1,\ldots,p_n) \Psi_{in}) \tag{2.67}$$

(see (2.31)) is given by (2.32) - (2.34), where $\tilde{G}_\mu$ must be inserted in (2.32) instead of $\tilde{R}$, and $\tilde{g}^+_\mu(P;Q,-Q')$ instead of $\tilde{r}^{amp}(\ldots)$ in (2.34). The superscript "amp" is no longer necessary because we are now in the totally amputated case.

The reduction formulae remain valid under the simultaneous replacement of "in" by "out" and of $\tilde{g}^+_\mu$ by $\tilde{g}^-_\mu$.

By taking the vacuum expectation value of (2.64) and summing on the right over complete sets of intermediate in-states we obtain an expression for $\tilde{g}_\mu$ in terms of integrals over products of $\tilde{r}$'s. Assuming property iii) of Theorem 2.1 for $\tilde{r}$ we deduce from this expression that $\tilde{g}_\mu$ also has this property iii).

From (2.61) we obtain the completeness equations (generalised GLZ equations)

$$g_\mu(X) - g_\nu(X) = -i \sum_{\ell=1}^{\infty} \frac{1}{\ell!} \int \prod_1^\ell \{du_i\, dv_i\}\, K^\ell(U-V)\, g^+_\alpha(X_L;U)\, g^+_\beta(X_R;V) \tag{2.68}$$

with $X_L = \{x_{i_1},\ldots,x_{i_k}\}$, $X_R = \{x_{i_{k+1}},\ldots,x_{i_n}\}$. For the definition of K^ℓ see (2.37). Eq. (2.68) remains true if the two factors $g^+_\cdot$ on the right are replaced by the corresponding $g^-_\cdot$.

The expansion (2.45) can be further generalised to

$$G_\mu(X) = g_\mu(X) + \sum_{\ell=1}^{\infty} \frac{1}{\ell!} \int du_1 \ldots du_\ell\, g^+_\mu(X;U) :A^{in}(u_1)\ldots A^{in}(u_\ell): \tag{2.69}$$

The <u>time ordered product</u> will not be used in our formalism. Since it is important in the applications, due to its close relation to the S-matrix, we must nevertheless mention its definition and major properties.

The T-product $T(x_1,\ldots,x_n)$ of the fields $A(x_1),\ldots, A(x_n)$ can be defined recursively by the formula

$$R(X) = T(X) + i \sum_{L,R} T^*(X_L)\, T(x_1, X_R) \quad , \tag{2.70}$$

the sum extending over all partitions of $\{x_2,..,x_n\}$ into two complementary subsets X_L and X_R , where X_R may be empty, X_L not*. $T(X)$ is again an operator-valued tempered distribution possessing properties i), ii), and vii) of the definition of R . The vacuum expectation value $\tau(x_1,..,x_n)$ of $T(x_1,..,x_n)$ is called a <u>time ordered function</u> or <u>Green's function</u>. $\tau(X)$ can be expanded in a cluster sum:

$$\tau(x_1,..,x_n) = \sum \tau^T(x_1,..,x_i) \ldots \tau^T(x_i,..,x_i) \quad , \tag{2.71}$$

the sum extending over all partitions of $\{x_1,..,x_n\}$ into any number of subsets with at least two elements. The <u>truncated Green's functions</u> are intimately connected with the retarded functions: they can be obtained from r by analytic continuation in p-space. Under our assumptions on r it can be proved that $\tilde{\tau}^T(p_1^-, \underset{\sim}{p}_1, \ldots, p_n^-, \underset{\sim}{p}_n, q_1^+, \underset{\sim}{q}_1, \ldots, q_m^+, \underset{\sim}{q}_m)$ is after integration over a test function in $\underset{\sim}{P}$, $\underset{\sim}{Q}$, continuous in P^-, Q^+ , in a neighbourhood of the mass shell $p_i^- = q_j^+ = 0$. Its value on the mass shell is, apart from a numerical factor, the connected part of the S-matrix element for the process: m incoming particles going into n outgoing particles.

Finally we must briefly discuss the <u>CTP properties</u> of the objects of our theory. Since the charge conjugation C is the identity in our theory, CTP is the same as TP. According to the CTP theorem [13,14] there exists in $\mathfrak{H}$ an anti-unitary operator Θ with the following properties:

$$\Theta^2 = 1 \quad , \tag{2.72}$$

$$\Theta\Omega = \Omega \quad , \tag{2.73}$$

$$\Theta A(x) \Theta = A(-x) \quad , \tag{2.74}$$

$$\Theta A^{in}(x) \Theta = A^{out}(-x) \quad . \tag{2.75}$$

---

* This definition of T seems rather roundabout. For another approach which defines T <u>first</u> in a natural way, and <u>then</u> R and G in terms of T , see Ref. [17] .

Let $\bar{C}_\mu$ be the n-cell obtained from C_μ by reversing all the signs σ_I , $\bar{G}_\mu$ the corresponding g.r.p. Then

$$\Theta\, G_\mu(x_1,..,x_n)\, \Theta = \bar{G}_\mu(-x_1,..,-x_n) \quad . \tag{2.76}$$

In particular, the retarded product $R(X)$ is transformed by Θ into the advanced product $\bar{R}(-X)$.

Finally:

$$\Theta\, T(x_1,..,x_n)\, \Theta = T^*(-x_1,..,-x_n) \quad . \tag{2.77}$$

For the vacuum expectation values of G_μ and T we have the CTP relations

$$g_\mu(x_1,..,x_n) = \bar{g}_\mu(-x_1,..,-x_n) \quad , \tag{2.78}$$

$$\tau(x_1,..,x_n) = \tau^*(-x_1,..,-x_n) \quad . \tag{2.79}$$

III. PERTURBATION THEORY: THE BASIC FORMALISM

We consider the theory of a family of fields $A(x,g)$ depending on a parameter g , the coupling constant, in such a way that $A(x) = A(x,g)$ for g fixed has the properties described in Chapter II. $A(x,0)$ shall be a free field of mass m . (The case that A depends on several coupling constants can be treated analogously.) We assume that any quantity $\mathcal{Q}$ of the theory (field operators, R-operators, r-functions, etc.) can be developed into a power series in g :

$$\mathcal{Q}(g) = \sum_{\sigma=0}^{\infty} \mathcal{Q}_\sigma\, g^\sigma \quad , \tag{3.1}$$

where $\mathcal{Q}_0$ is the value of the respective quantity in the free case. The problem of the convergence of the series (3.1) will not be touched, since we cannot say more about it than in the canonical approach: nothing. We are only concerned with the existence and calculation of the expansion coefficients $\mathcal{G}_\sigma$ in arbitrary finite orders σ . (3.1) will, then, be treated as a formal power series. It is hoped that it is at least an asymptotic expansion for $g \to 0$.

$A_o(x)$ is a free field with the conventions introduced in Chapter II for A^{ex} , so that we have

$$A_o^{in}(x) = A_o^{out}(x) = A_o(x) \quad . \tag{3.2}$$

The free values of the retarded functions are

$$\begin{aligned} r_o(x,y) &= -\,(\Box_y + m^2)\,\delta^4(x-y) \quad , \\ \tilde{r}_o(p,q) &= \frac{1}{2\pi}\,(q^2-m^2)\,\delta^4(p+q) \quad , \end{aligned} \tag{3.3}$$

$$r_o(x_1,..,x_n) = 0 \qquad \text{for} \quad n > 2 \quad , \tag{3.4}$$

and

$$g_{\mu o}(x_1,..,x_n) = r_o(x_1,..,x_n) \tag{3.5}$$

for all μ .

In order to find the fields $A_\sigma(x)$ we use the GLZ theorem. We try to find solutions of the amputated GLZ equations satisfying the subsidiary conditions stated in Theorems 2.1 and 2.2 .

We introduce the following abbreviations:

$$J_{\sigma L}(x,y,X) = \int \prod_1^\ell \{\, du_i\, dv_i \;\; \Delta_+(u_i-v_i) \,\}\; r(x,X_L,U)\; r(y,X_R,V) \quad , \tag{3.6}$$

with $U = \{u_1,..,u_\ell\}$, $V = \{v_1,..,v_\ell\}$, $X = \{x_1,..,x_n\}$, and X_L , X_R defined as in (2.36),

$$I(x,y,X) = -i \sum_{L,R} \sum_{\ell=1}^{\infty} \frac{i^\ell}{\ell!} \left[J_{\ell L}(x,y,X) - J_{\ell L}(y,x,X) \right] \quad , \tag{3.7}$$

the right-hand side of (2.36). The GLZ equations read then

$$r(x,y,X) - r(y,x,X) = I(x,y,X) \quad . \tag{3.8}$$

These equations we must solve in perturbation theory, with the subsidiary conditions mentioned above, and with boundary conditions specifying the dynamics that will be introduced later. We substitute the expansion (3.1) of r into (3.8) and equate the terms of equal order g^σ on both sides:

$$r_\sigma(x,y,X) - r_\sigma(y,x,X) = I_\sigma(x,y,X) \quad , \tag{3.9}$$

$$I_\sigma(x,y,X) = -i \sum_{L,R} \sum_{\ell=1}^{\infty} \frac{i^\ell}{\ell!} \left[J_{\ell L\sigma}(x,y,X) - J_{\ell L\sigma}(y,x,X) \right] \quad , \tag{3.10}$$

$$J_{\ell L\sigma}(x,y,X) = \sum_{\tau=1}^{\sigma-1} \int \prod_1^\ell \{du_i\, dv_i \quad \Delta_+(u_i - v_i)\}\, r_\tau(x,X_L,U)\, r_{\sigma-\tau}(y,X_R,V) \quad . \tag{3.11}$$

That the sum in (3.11) does not contain terms with $\tau=0$ or σ is due to (3.4) and the non-occurrence of the 2-point function in I_σ (see the remarks preceding Theorem 2.2). In this fact lies the advantage of the amputated GLZ equations over their non-amputated form. It suggests immediately a recursive procedure.

Assume that the r_τ with $\tau<\sigma$ are known. Then we can calculate I_σ by simple integration, and r_σ can be determined as solution of the linear equation (3.9) whose right-hand side is known. This program we shall carry out in what follows. It can be divided into two phases.

Phase 1: formal solution. We construct, by recursion, formal solutions of (3.9) satisfying conditions i) and ii) of Theorem 2.1 and the normalisation condition (2.58), without worrying about existence problems. Condition iii) of Theorem 2.1. will be disregarded in this phase.

Phase 2: existence. The existence of the formal expressions derived in phase 1 is established. Here condition iii) becomes essential.

The word "formal" as used here should not be understood in too naive a sense. At first sight one might be inclined to consider the problem of solving (3.9) trivial. $r_\sigma(x,y,X)$ has its support in $(x-y)\in\overline{V}_+$, $r_\sigma(y,x,X)$ in $(y-x)\in\overline{V}_+$, and these two supports overlap only in the points $x \neq y$. Hence

$$r_\sigma(x,y,X) = \theta(x-y)\, I_\sigma(x,y,X)$$

seems to be the obvious solution. In reality things are not nearly that simple. The complication comes from the fact that I_σ is a distribution, not a function, and we look for solutions r_σ that are also distributions. This means that, firstly, the product θI_σ is not defined a priori and, secondly, that the manifold $x=y$ cannot be ignored despite its low dimension.

Chapters IV-VI will deal with phase 1, Chapter VII with phase 2.

IV. THE EQUATION $r_\sigma(x,y,X) - r_\sigma(y,x,X) = I_\sigma(x,y,X)$

In this chapter the solutions of (3.9) satisfying conditions i) and ii) of Theorem 2.1 will be discussed. We start with the

Problem of uniqueness. Eq. (3.9) is a linear equation. Its solutions are therefore unique up to solutions of the homogeneous equation

$$h(x,y,x_1,..,x_n) - h(y,x,x_1,..,x_n) = 0 \quad , \qquad (4.1)$$

where h has to satisfy the invariance and support conditions demanded for r_σ .

The most general solution of (4.1), disregarding the subsidiary conditions , is a tempered distribution which is symmetric in the first two variables x, y . The condition i) of symmetry in $\{y,x_1,..,x_n\}$ implies then that h is invariant under permutations of all its arguments. The support condition ii) demands support in $(x-y)\in\bar{V}_+$, and because of the (x,y) symmetry also in $(y-x)\in\bar{V}_+$, hence in $x=y$. Because of the total symmetry just derived we obtain

$$\text{supp } h \subset \{x=y=x_1=\ldots=x_n\} \quad . \tag{4.2}$$

Because of translation invariance h depends only on the variables $\xi = x-y$, $\xi_i = x-x_i$, and has its support in the point $\xi = \xi_i = 0$. The only distributions with this support are of the form (see Ref. [19] , p.100)

$$D\ \delta^4(\xi) \prod_{i=1}^{n} \delta^4(\xi_i) \quad ,$$

with D an arbitrary differential operator with the appropriate symmetry. Finally, because of the rest of condition i), D must be real and Lorentz invariant. As a result we obtain

<u>Theorem 4.1</u>

The most general solution of equation (4.1) satisfying conditions i) and ii) of Theorem 2.1 is

$$h(x,y,x_1,..,x_n) = D\ \delta^4(x-y) \prod_{i=1}^{n} \delta^4(x-x_i) \quad , \tag{4.3}$$

where D is a differential operator in x, y, X, with constant, real coefficients, which is Lorentz invariant and totally symmetric in x, y, X . In p-space this becomes

$$\tilde{h}(p,q,p_1,..,p_n) = \delta^4(p+q+\ \ p_i)\ \mathcal{P}(p,q,p_1,..,p_n) \tag{4.4}$$

with $\mathcal{P}$ an invariant, real, totally symmetric polynomial.

As an important application of this theorem we consider the first order ($\sigma = 1$) form of (3.9). Obviously:

$$I_1(x,y,X) = 0 \quad , \tag{4.5}$$

hence r_1 is of the form (4.3). Within the restrictions given in Theorem 4.1..we can choose the differential operators D freely. <u>It is the choice of these D that fixes the interaction</u>, i.e. the choice of the D corresponds to the choice of an interaction Lagrangian (or

Hamiltonian) in canonical field theory.

For simplicity we choose all the D except one as zero, so that only one r_1 is different from zero. Let this be the μ-point function, with $\mu > 2$: *

$$r_1(x_1,..x_n) = 0 \quad \text{for} \quad n \neq \mu \ ,$$
$$r_1(x_1,..,x_\mu) = D \prod_{i=1}^{\mu} \delta^4(x_1 - x_i) \ . \tag{4.6}$$

Furthermore, we choose D to be homogeneous of order ν , so that the polynomial $\mathcal{P}$ in (4.4) is a form of degree ν . Such a theory will be called <u>of type (μ,ν)</u> .

The interaction Lagrangian corresponding to (4.6) is

$$L_{int} = \frac{g}{\mu!} \int dx_1..dx_\mu \; r_1(x_1,..,x_\mu) \; :A(x_1)..A(x_\mu): \ , \tag{4.7}$$

i.e. the r_1 calculated from this L_{int} in the canonical way of doing things are given by (4.6). L_{int} is of degree μ in the fields and contains derivatives of order ν . Note that high derivatives in r_1 lead to high derivatives in L_{int} , in which case the canonical formalism encounters the well-known difficulties. In contrast to this, the formalism to be developed here has no difficulty accomodating arbitrarily high derivatives.

The theories of type (3,0) and (4,0) correspond to the two so-called renormalisable theories of the canonical formalism, the A^3- and the A^4- theory respectively. For the meaning of renormalisability in our formalism see Chapter VIII.

The choice of r_1 does, of course, not fix the theory uniquely, since ambiguities of the form (4.3) occur in each order. We shall see in Chapter V how one is to deal with these ambiguities in orders $\sigma > 2$.

---

* The choice $\mu=2$ with D of order 0 or 2 would destroy the normalisation condition (2.58): it would correspond to a renormalisation of mass or wave function respectively. D's of higher order (still for $\mu=2$) are uninteresting because of the limited part that the 2-point function plays in our formalism.

We come now to the

<u>Problem of existence</u>. We prove

<u>Theorem 4.2.</u>

In order that the equation

$$r_\sigma(x,y,x_1,..,x_n) - r_\sigma(y,x,x_1,..,x_n) = I_\sigma(x,y,x_1,..,x_n) \qquad (3.9)$$

has a solution $r_\sigma(x,y,X)$ satisfying conditions i) and ii) of Theorem 2.1 , it is necessary and sufficient that $I_\sigma(x,y,X)$ is a tempered distribution with the following properties.

a) supp $I_\sigma(x,y,X)$ $\{ (x-y)^2 \geq 0 \; ; \; (x-x_i)\in\bar{V}_+$ or $(y-x_i)\in\bar{V}_+$ for all $i=1,...,n \}$. (4.8)

b) I_σ satisfies the identities

$$I_\sigma(x,y,X) + I_\sigma(y,x,X) = 0 \quad , \qquad (4.9)$$

$$I_\sigma(x,y,z,X) + I_\sigma(y,z,x,X) + I_\sigma(z,x,y,X) = 0 \quad . \qquad (4.10)$$

c) I_σ is real.

d) I_σ is invariant under the connected Poincaré group $\mathcal{P}_+^\uparrow$.

e) $I_\sigma(x,y,X)$ is symmetric in the arguments $X = \{x_1,..,x_n\}$.

Conditions (4.10) and e) are absent in the 2-point case.

It is trivial to see that the stated conditions are necessary. A remark with respect to the support condition a) is in order. From the desired support of r_σ we obtain as a necessary condition for solvability of (3.9) that I_σ vanish if not all the x_i are retarded with respect to x , or all x_i retarded with respect to y . At first sight this seems to be stronger than the second part of (4.8), where we demand only that each x_i be retarded with respect to either x or y , the choice among the two depending on i . This, together with the condition $(x-y)^2 \geq 0$, implies however the stronger form just given. If $(x-y)\in\bar{V}_+$, then all the x_i retarded with respect to y are also retarded with respect to x , and if $(y-x)\in\bar{V}_+$, then all x_i retarded with respect to either x or y are retarded with respect to y .

That the conditions a) - e) are sufficient will now be shown by explicit construction of a solution. We want r_σ to be a tempered distribution, i.e. a continuous linear form over the space $\mathcal{S}$ of strongly decreasing test functions [19], and as such we will construct the solution. The construction will be carried out in different steps, successively enlarging the spaces of test functions on which r_σ is defined. Before starting, let us introduce a piece of notation: we say that the test function $\phi(x_1,..,x_n)$ "vanishes strongly" at the point $(a_1,..,a_n)$, written $\phi(a_1,..,a_n) \equiv 0$, if ϕ vanishes at this point together with all its derivatives.

<u>1st step</u>. Let $\phi(x,y,x_1,..,x_n) \in \mathcal{S}$, with $\phi \equiv 0$ at $x=y$. We introduce an auxiliary function $\vartheta(\xi)$ of the 4-vector ξ with the following properties.

α) $\vartheta(\xi)$ is real.

β) $\vartheta(\xi)$ is scale invariant:

$$\vartheta(\lambda\xi) = \vartheta(\xi) \quad \text{for } \lambda > 0 \ . \tag{4.11}$$

γ) $\vartheta(\xi)$ is C^∞ everywhere except at $\xi = 0$.

δ)
$$\vartheta(\xi) = \begin{cases} 1 & \text{for } \xi \in V_+ \\ 0 & \text{for } \xi \in V_- \ . \end{cases} \tag{4.12}$$

Under these conditions we have

$$|D\,\vartheta(\xi)| < c_D \ |\xi|^{-|D|} \tag{4.13}$$

for D any derivation of order $|D|$, where $|\xi|$ is the Euclidean length of ξ , and c_D is a suitable positive constant. The strong vanishing of ϕ at $x=y$ implies the existence of constants C_M^D , $M=0,1,2,...$, such that

$$|D\phi| < C_M^D \ |x-y|^M \ . \tag{4.14}$$

This implies that the product $\vartheta(x-y)\phi(x,y,X)$ is in $\mathcal{S}$, the singularity of ϑ at $x=y$ being cancelled by the vanishing of ϕ .

We define*

$$\langle r_\sigma | \phi \rangle \quad = \quad \langle I_\sigma | \vartheta\phi \rangle \quad . \tag{4.15}$$

This expression exists and does indeed solve (3.9):

$$\langle r_\sigma(x,y,X) - r_\sigma(y,x,X) \mid \phi(x,y,X) \rangle$$

$$= \langle I_\sigma(x,y,X) \mid \vartheta(x-y)\ \phi(x,y,X) \rangle \ - \ \langle I_\sigma(y,x,X) \mid \vartheta(y-x)\ \phi(x,y,X) \rangle$$

$$= \langle I_\sigma(x,y,X) \mid \phi(x,y,X) \rangle \quad .$$

We have used (4.9) and the fact that $\vartheta(x-y) + \vartheta(y-x) = 1$ in supp I_σ. The indeterminacy of this sum at $x=y$ is unimportant because ϕ vanishes there.

The definition (4.15) does not depend on the choice of ϑ, because a freedom of choice exists only in $(x-y)^2 < 0$, and there I_σ vanishes.

It is easy to see that r_σ as defined in (4.15) has the correct support: it vanishes on test functions ϕ which vanish in $(x-y) \epsilon \overline{V}_+$, $(x-x_i) \epsilon \overline{V}_+$ for all i.

2nd step. Choose an arbitrary x_i and rename it z. Let $\phi(x,y,..,z,..) \epsilon \mathscr{S}$ vanish strongly for $x=z$. Due to the symmetry requirement for r_σ we obtain from (4.15):

$$\langle r_\sigma(x,y,..,z,..) \mid \phi(x,y,..,z,..) \rangle$$

$$= \langle I_\sigma(x,z,y,...) \mid \vartheta(x-z)\ \phi(x,y,..,z,..) \rangle \quad . \tag{4.16}$$

We must show that this definition agrees with (4.15) on test functions ϕ which vanish strongly both for $x=y$ and for $x=z$. The variables x_i different from z are irrelevant for this problem. For ease of

---

* For the value of the distribution T on the test function ϕ we use alternatively the symbols $\langle T|\phi \rangle$, $\langle T(X)|\phi(X) \rangle$, or $\int dX\ T(X)\phi(X)$. We apologise to the reader for this low-brow notation.

notation we shall therefore treat only the 3-point case. The generalisation to arbitrary n is immediate. Let $\phi(x,y,z)$ be a test function with $\phi \equiv 0$ at $x=y$ and at $x=z$. Consider

$$A = \langle I_\sigma(x,y,z) | \vartheta(x-y)\ \phi(x,y,z) \rangle \quad .$$

$\vartheta(x-y)$ vanishes in $(x-y) \in V_-$ by definition. In view of (4.8) only the points $(x-y) \in \bar{V}_+$ contribute then to A . The second part of (4.8) yields the additional requirement $(x-z) \in \bar{V}_+$. But there $\vartheta(x-z) = 1$, so that A can be written

$$A = \langle I_\sigma(x,y,z) | \vartheta(x-y)\ \vartheta(x-z)\ \phi(x,y,z) \rangle \quad .$$

With (4.10) this becomes

$$A = - \langle I_\sigma(y,z,x) | \vartheta(x-y)\ \vartheta(x-z)\ \phi(x,y,z) \rangle$$
$$- \langle I_\sigma(z,x,y) | \vartheta(x-y)\ \vartheta(x-z)\ \phi(x,y,z) \rangle \quad .$$

The first term vanishes because the supports of the two factors do not overlap. In supp $I_\sigma(y,z,x)$ we have $(y-x) \in \bar{V}_+$ or $(z-x) \in \bar{V}_+$, in either case one of the ϑ-functions vanishes. To the second term we can add the expression

$$- \langle I_\sigma(z,x,y) | \vartheta(y-x)\ \vartheta(x-z) \phi(x,y,z) \rangle$$

which vanishes for a similar reason ($(x-z) \in \bar{V}_+$, and $(z-y) \in \bar{V}_+$ or $(x-y) \in \bar{V}_+$, implies $(x-y) \in \bar{V}_+$), so that we obtain finally by using (4.9)

$$A = \langle I_\sigma(x,z,y) | \vartheta(x-z)\ \phi(x,y,z) \rangle \quad ,$$

the desired consistency relation.

<u>3rd step</u>. Let $\mathcal{S}_0 \subset \mathcal{S}$ be the space of test functions

$\phi(x,x_1,..,x_n)$ which vanish strongly for $x=x_1=...=x_n$.

Let $f(u)$ be a real C^∞-function which vanishes strongly at $u = 0$ (u is a scalar variable) and is strictly positive for $u\neq 0$. Define

$$F(u_1,...,u_n) = \sum_{i=1}^{n} f(u_i) \quad .$$

F is C^∞ and strictly positive outside the origin, in particular on the unit sphere $\Sigma u_i^2 = 1$. Let $\xi_i = x-x_i$, $|\xi_i|^2 = (\xi_i^o)^2 + \underline{\xi}_i^2$, $|\Xi|^2 = \Sigma_{i=1}^n |\xi_i|^2$, and $u_i = |\xi_i| / |\Xi|$. We introduce the auxiliary functions

$$f_i(x,x_1,..,x_n) = \frac{f(u_i)}{F(u_1,..,u_n)} \quad . \tag{4.17}$$

f_i is real, C^∞ everywhere except for $x=x_1=..=x_n$, invariant under translations and dilatations, and vanishes strongly at $x=x_1$. Furthermore:

$$\sum_{i=1}^{n} f_i(x,...,x_n) = 1 \quad . \tag{4.18}$$

Because of the scaling invariance of f_i we have

$$|D f_i| \leq c_D \ |\Xi|^{-|D|} \tag{4.19}$$

in analogy to (4.13). In analogy to (4.14) we have for $\phi \in \mathcal{S}_o$:

$$|D\phi| \leq C_M^D \ |\Xi|^M \quad , \tag{4.20}$$

hence the product $\phi_i = f_i \ \phi$ is in $\mathcal{S}_o$. Moreover ϕ_i vanishes strongly at $x=x_i$ so that the value of r_σ on ϕ_i is defined by the results of the 2nd step. Because of

$$\phi(x,X) = \sum_{i=1}^{n} \phi_i(x,X) \tag{4.21}$$

we obtain

$$\langle r_\sigma(x,X)\,|\,\phi(x,X)\rangle = \sum_{i=1}^{n} \langle I_{i\sigma}(x,X)\,|\,\zeta_i(x,X)\;\phi(x,X)\rangle , \quad (4.22)$$

with

$$I_{i\sigma}(x,x_1,..,x_n) = I_\sigma(x,x_i,x_1,..,x_{i-1},x_{i+1},..,x_n) , \quad (4.23)$$

$$\zeta_i(x,X) = \vartheta(x-x_i)\; f_i(x,X) . \quad (4.24)$$

The expression (4.22) exists and does not depend on the special choice of f, as can be seen by a simple adaptation of the arguments of the 2nd step.

(4.22) defines r_σ on $\mathcal{S}_0$ with the correct properties. Symmetry in the x_i, reality, and translational invariance are trivially satisfied. The support in $(x-x_i)\in\overline{V}_+$ follows from the supports of ϑ and I_σ (our construction procedure has been specially geared to satisfy the support condition). The only condition that still must be checked is Lorentz invariance. Let Λ be a proper Lorentz transformation. $\vartheta(\Lambda\xi)$ and $f_i(\Lambda x,\Lambda X)$ have all the relevant properties of the auxiliary functions ϑ and f_i and can therefore replace these functions in (4.22) without changing the result. Hence

$$\langle r_\sigma(x,X)\,|\,\phi(\Lambda x,\Lambda X)\rangle = \sum_i \langle I_{i\sigma}(x,X)\,|\,\zeta_i(x,X)\;\phi(\Lambda x,\Lambda X)\rangle$$

$$= \sum_i \langle I_{i\sigma}(x,X)\,|\,\zeta_i(\Lambda x,\Lambda X)\;\phi(\Lambda x,\Lambda X)\rangle$$

$$= \sum_i \langle I_{i\sigma}(\Lambda^{-1}x,\Lambda^{-1}X)\,|\,\zeta_i(x,X)\;\phi(x,X)\rangle$$

$$= \sum_i \langle I_{i\sigma}(x,X)\,|\,\zeta_i(x,X)\;\phi(x,X)\rangle = \langle r_\sigma(x,X)\,|\,\phi(x,X)\rangle ,$$

q.e.d. We have used the invariance of I_σ and the relation

$$\langle T(X)\,|\,\phi(X)\rangle = \langle T(\Lambda X)\,|\,\phi(\Lambda X)\rangle$$

which is valid for any distribution T . In fact, this is the definition of the action of Λ on T .

4th step. We must still extend the definition of r_σ from $\mathcal{S}_o$ to the full space $\mathcal{S}$. This we do by extending it to an even larger space of test functions.

Let $\mathcal{S}^N$ be the space of functions $\phi(x,x_1,\ldots,x_n)$ which are N times continuously differentiable and for which

$$\sup_{x,X} \left| (1 + |x|^2 + |X|^2)^\alpha \; D\phi(x,X) \right| < \infty \tag{4.25}$$

for all non-negative integers α and all derivatives D of order $|D| \leq N$. The topolgy of $\mathcal{S}^N$ is defined by the seminorms (4.25). $\mathcal{S}$ is dense in $\mathcal{S}^N$. According to a well-known theorem (ref. [19] p.239) there exists, for a given I_σ , a positive integer N such that I_σ is defined on $\mathcal{S}^N$. We shall define r_σ on the same $\mathcal{S}^N$.

Let $\mathcal{S}_o^N$ be the subspace of $\mathcal{S}^N$ consisting of the test functions which vanish with all derivatives up to order N at the points $x=x_1=\ldots=x_n$. It is easy to see that all the arguments of steps 1-3 are still valid if the strong vanishing of ϕ in a point is defined in this new restricted way. The definition (4.22) can be extended immediately to $\mathcal{S}_o^N$.

We remember that, due to translational invariance, $I_\sigma(x,X)$ and $r_\sigma(x,X)$ can be considered as distributions in the difference variables $\xi_i = x-x_i$, $i=1,\ldots,n$, (see Ref. [13] pp. 38-40). $r_\sigma(\xi_1,\ldots,\xi_n)$ is defined by the previous considerations on the space $\mathcal{S}_o^N(\Xi)$ of N times differentiable test functions $\phi(\Xi)$ which vanish strongly at $\xi_1=\ldots=\xi_n=0$. We must extend this definition to $\mathcal{S}^N(\Xi)$.

Let D denote the differential operator

$$D = \prod_{i=1}^{n} \prod_{\nu=o}^{3} \frac{\partial^{\alpha_{i\nu}}}{(\partial\xi_i^\nu)^{\alpha_{i\nu}}} \tag{4.26}$$

$|D| = \sum_{i,\nu} \alpha_{i\nu}$ its order. Define

$$\Xi^D = \prod_{i,\nu} \frac{(\xi_i^\nu)^{\alpha_{i\nu}}}{\alpha_{i\nu}!} \quad . \tag{4.27}$$

Let $\gamma(\xi_1,..,\xi_n)$ be an arbitrary but fixed real C^∞-function with compact support, such that $\gamma \equiv 1$ in a neighbourhood of the origin and such that $\gamma(x-x_1,...,x-x_n)$ is invariant under permutations of the arguments $x,x_1,..,x_n$. Such functions exist. For example, we can form

$$\gamma(\Xi) = \prod_{i=1}^{n} \bar{\gamma}(x-x_i) \prod_{i<j} \bar{\gamma}(x_i-x_j)$$

with $\bar{\gamma}(\xi)$ of compact support and $\equiv 1$ in a neighbourhood of $\xi = 0$. The differences in the arguments of $\bar{\gamma}$ have to be expressed in terms of ξ_i .

Define

$$\gamma^D(\Xi) = \Xi^D \gamma(\Xi) \tag{4.28}$$

for all D with $|D| \leqslant N$. For D, D' two operators of the form (4.26) we have

$$D' \gamma^D \Big|_{\xi_1=..=\xi_n=0} = \delta_{DD'} \quad , \tag{4.29}$$

where equality of two differential operators is defined as equality of all the exponents $\alpha_{i\nu}$.

Let $\phi(\Xi) \in \mathcal{S}^N$. We can expand

$$\phi(\Xi) = \sum_{\substack{D \\ |D| \leqslant N}} \phi_D \gamma^D(\Xi) + \phi'(\Xi) \quad , \tag{4.30}$$

with

$$\phi_D = D \phi(\Xi) \Big|_{\xi_1=..=\xi_n=0} \tag{4.31}$$

and $\phi'(\Xi) \in \mathcal{S}_o^N$. Hence

$$\langle r_\sigma(\Xi)|\phi(\Xi)\rangle = \langle r_\sigma(\Xi)|\phi'(\Xi)\rangle + \sum_D \phi_D \langle r_\sigma(\Xi)|\gamma^D(\Xi)\rangle \quad . \tag{4.32}$$

The first term on the right is defined by the results of the 3rd step. We must determine the finitely many constants $\langle r_\sigma|\gamma^D\rangle$ such that r_σ satisfies all the necessary requirements.

We can write $\phi(\Xi)$ as a translation invariant function of x, X, by substituting $\xi_i = x - x_i$ in the argument. The conditions on r_σ are then:

1) Reality. This is satisfied if $\langle r_\sigma|\gamma^D\rangle$ is real.

2) Support. This is automatically satisfied by (4.32) since a ϕ that vanishes in T_n has vanishing ϕ_D .

3) Symmetry in X . If $\phi(x,..,x_i,..,x_j,..) = -\phi(x,..,x_j,..,x_i,..)$ for any pair (i,j) , then

$$\langle r_\sigma|\phi\rangle = 0 \quad . \tag{4.33}$$

4) The equation (3.9) must be satisfied: if $\phi(x,x_1,...) = -\phi(x_1,x,...)$ then*

$$2\langle r_\sigma(x,x_1,...)|\phi(x,x_1,...)\rangle$$

$$= \langle r_\sigma(x,x_1,...)|\phi(x,x_1,...)\rangle - \langle r_\sigma(x_1,x,...)|\phi(x,x_1,...)\rangle$$

$$= \langle I_\sigma(x,x_1,...)|\phi(x,x_1,...)\rangle \quad . \tag{4.34}$$

One convinces himself easily that the definitions (4.33), (4.34) coincide for ϕ which are antisymmetric in two of variables. For example, assume $\phi(x,y,z) = -\phi(y,x,z) = -\phi(x,z,y)$, so that ϕ is

---

* The expression $\langle r_\sigma(x,...,x_n)|\phi(x,..,x_n)\rangle$ must here be read as $\langle r_\sigma(\xi_1,..,\xi_n)|\phi(\xi_1,..,\xi_n)\rangle$ since $\phi(x,...,x_n)$ is not a test function.

invariant under cyclic permutations of the 3 variables. Then

$$\tfrac{1}{2}\langle I_\sigma(x,y,z)\,|\,\phi(x,y,z)\rangle = \tfrac{1}{6}\langle I_\sigma(x,y,z)\,|\,\phi(x,y,z) + \phi(y,z,x) + \phi(z,x,y)\rangle$$

$$= \tfrac{1}{6}\langle I_\sigma(x,y,z) + \text{cycl.}\,|\,\phi(x,y,z)\rangle$$

$$= 0$$

because of (4.10). Hence the definitions (4.33) and (4.34) give the same result.

We introduce the finite dimensional vector space Γ spanned by the functions γ^D written as functions of the variables $x, x_1, \ldots, x_n$. A linear representation of the permutation group $\mathcal{S}_{n+1}$ of $n+1$ objects is defined on this space in an obvious way through permutations of the variables x, x_i . This representation can be reduced with the help of Young tableaux*, and we obtain

$$\Gamma = \Gamma_s \oplus \Gamma_a \qquad (4.35)$$

with Γ_s consisting of totally symmetric functions, while the functions in Γ_a can be written as a sum over terms each of which is antisymmetric in at least one pair of variables. (4.33) and (4.34) define r_σ on Γ_a but not on Γ_s .

We rewrite the sum in (4.30) as

$$\sum_D \phi_D\, \gamma^D(\Xi) = \gamma_s(\phi;\Xi) + \gamma_a(\phi;\Xi) \quad , \qquad (4.36)$$

$\gamma_s \in \Gamma_s$, $\gamma_a \in \Gamma_a$ as functions of x, X . γ_s and γ_a depend on ϕ only through the constants ϕ_D , $|D| \leq N$. $\langle r_\sigma|\gamma_a\rangle$ is fixed uniquely, $\langle r_\sigma|\gamma_s\rangle$ is still free.

---

* See any of the numerous books on the applications of group theory in quantum mechanics.

We must yet satisfy a fifth requirement: Lorentz invariance. This will necessitate a somewhat lengthy consideration.*

$r_\sigma(x-x_1,\ldots,x-x_n)$ is invariant under the connected Lorentz group if it is invariant under infinitesimal Lorentz transformations. It is therefore sufficient to satisfy the conditions

$$M_{\mu\nu} r_\sigma(x-x_1,\ldots,x-x_n) = 0 \quad , \tag{4.37}$$

with

$$M_{\mu\nu} = \left(x^\mu \frac{\partial}{\partial x^\nu} \pm x^\nu \frac{\partial}{\partial x^\mu}\right) + \sum_{i=1}^{n} \left(x_i^\mu \frac{\partial}{\partial x_i^\nu} \pm x_i^\nu \frac{\partial}{\partial x_i^\mu}\right) \tag{4.38}$$

for $\mu,\nu = 0,\ldots,3$, $\mu<\nu$. The upper sign holds for $\mu=0$, the lower sign in the other cases.

The action of $M_{\mu\nu}$ on functions $\phi(\xi_1,\ldots,\xi_n)$ is defined with the help of the substitution $\xi_i = x-x_i$. $M_{\mu\nu}$ maps $\mathcal{S}^{N+1}$ into $\mathcal{S}^N$. In distribution terms (4.37) means that $r_\sigma(\Xi)$ vanishes on all test functions of the form $M_{\mu\nu}\psi(\Xi)$, $\psi \in \mathcal{S}^{N+1}$.

Let $\mathcal{V}_M$ be the finite dimensional vector space of all forms of degree M in the variables ξ_i . The vectors Ξ^D of (4.27) with $|D| = M$ form a basis of $\mathcal{V}_M$. A representation of the permutation group $\mathcal{T}_{n+1}$ is defined on $\mathcal{V}_M$ in the same way as it was on Γ . Therefore we can again split $\mathcal{V}_M$:

$$\mathcal{V}_M = \mathcal{V}_M^s \oplus \mathcal{V}_M^a \quad , \tag{4.39}$$

$\mathcal{V}_M^s$ containing only totally symmetric forms, $\mathcal{V}_M^a$ the rest.

---

* A simpler solution of the invariance problem, based on the analyticity properties of the g.r.f. in p-space, has been developed by Epstein and Glaser in connection with their new approach to perturbation theory [16,17] .

$M_{\mu\nu}$ operates linearly on $\mathcal{W}_M$. Because of its symmetry under permutations it maps $\mathcal{W}_M^s$ into $\mathcal{W}_M^s$, $\mathcal{W}_M^a$ into $\mathcal{W}_M^a$. We define a scalar product in $\mathcal{W}_M$ by demanding the basis

$$e_D = \prod_{i,\nu} \sqrt{\alpha_{i\nu}!}\ \Xi^D \tag{4.40}$$

to be orthonormal (see (4.27)). In this basis the M_{ok}, k=1,2,3, are represented by symmetrical matrices, the M_{ik}, i,k=1,2,3, by antisymmetrical matrices. The Casimir operator

$$C = \sum_{k=1}^{3} M_{ok}^2 - \sum_{1\leq i<k\leq 3} M_{ik}^2 \tag{4.41}$$

is hermitian positive and can be diagonalised. The eigenvectors v_i with eigenvalue 0 are annihilated by $M_{\mu\nu}$:

$$(v_i, Cv_i) = 0 \Longrightarrow \sum_{\mu<\nu} (M_{\mu\nu}v_i, M_{\mu\nu}v_i) = 0 \longrightarrow M_{\mu\nu}v_i = 0$$

since all the terms in $\Sigma_{\mu<\nu}$ are non-negative. Because of the total symmetry of $M_{\mu\nu}$ there exists a basis (not necessarily orthogonal) of eigenvectors of C all lying in $\mathcal{W}_M^s$ or $\mathcal{W}_M^a$. Let $\mathcal{W}_M^{so}$ be the subspace of $\mathcal{W}_M^s$ spanned by the C-eigenvectors with eigenvalue 0, $\mathcal{W}_M^{sl}$ its orthogonal complement in $\mathcal{W}_M^s$, spanned by the eigenvectors with eigenvalues $\neq 0$. Define $\mathcal{W}_M^{ao}$, $\mathcal{W}_M^{al}$ analogously and $\mathcal{W}_M^o = \mathcal{W}_M^{so} + \mathcal{W}_M^{ao}$, $\mathcal{W}_M^l = \mathcal{W}_M^{sl} + \mathcal{W}_M^{al}$. Thus

$$\mathcal{W}_M^s = \mathcal{W}_M^{so} \oplus \mathcal{W}_M^{sl}\ , \quad \mathcal{W}_M^a = \mathcal{W}_M^{ao} \oplus \mathcal{W}_M^{al}\ , \tag{4.42}$$

and

$$M_{\mu\nu} \, \mathcal{W}_M^{so} = 0 \ , \qquad M_{\mu\nu} \, \mathcal{W}_M^{ao} = 0 \ , \qquad (4.43)$$

$$C \, \mathcal{W}_M^{s1} = \mathcal{W}_M^{s1} \ , \qquad C \, \mathcal{W}_M^{a1} = \mathcal{W}_M^{a1} \ , \qquad (4.44)$$

i.e. C induces automorphisms of $\mathcal{W}_M^{s1}$, $\mathcal{W}_M^{a1}$, and hence of $\mathcal{W}_M^1$. The inverse C^{-1} exists as an automorphism on these 3 spaces.

With the help of these results we can rewrite the expansion (4.30) in a form which is better adapted to the problem at hand. Let again $\phi(\Xi) \in \mathcal{f}^N$. We can find forms $F_M(\Xi) \in \mathcal{W}_M$, $M \leqslant N$, such that we have in a neighbourhood of the origin

$$\phi(\Xi) = \sum_{M=0}^{N} F_M(\Xi) + R(\Xi) \ , \qquad (4.45)$$

with $R(\Xi)$ vanishing strongly at $\xi_i = 0$ (i.e. with all derivatives up to order N). The coefficients of F_M depend linearly on the ϕ_D , $|D| = M$. We split F_M into two parts:

$$F_M(\Xi) = F_M^o(\Xi) + F_M^1(\Xi) \ , \qquad F_M^\alpha \in \mathcal{W}_M^\alpha \ , \qquad (4.46)$$

and expand

$$\phi(\Xi) = \sum_{M=0}^{N} F_M^o(\Xi) \, \gamma(\Xi) + \sum_{M=0}^{N} C\left[\gamma(\Xi) \, C^{-1} F_M^1(\Xi)\right] + \phi''(\Xi) \ . \qquad (4.47)$$

C is defined by (4.41) , where the $M_{\mu\nu}$ are now the differential operators (4.38). The C^{-1} in ... is, however, still a matrix in $\mathcal{W}_M^1$. In the neighbourhood in which $\gamma \equiv 1$ the second term on the right of (4.47) is $\Sigma_M \, F_M^1(\Xi)$, hence $\phi'' \in \mathcal{f}_o^N$. Condition (4.37) implies that r_σ vanishes on the second sum in (4.47). F_M^o can be split further:

$$F_M^o(\Xi) = F_M^{so}(\Xi) + F_M^{ao}(\Xi) \quad , \quad F_M^{\alpha o} \epsilon \mathcal{W}_M^{\alpha o} \quad , \tag{4.48}$$

and we obtain finally

$$\langle r_\sigma | \phi \rangle$$

$$= \langle r_\sigma | \phi'' \rangle + \sum_{M=0}^{N} \langle r_\sigma | F_M^{ao} \gamma \rangle + \sum_{M=0}^{N} \langle r_\sigma | F_M^{so} \gamma \rangle \quad . \tag{4.49}$$

In this expression the first term (the ϕ''-term) is defined by the 3rd step, the second is defined by the 4th step because $F_M^{ao}\gamma \epsilon \Gamma_a$, but the third term remains undefined. It can be defined at will, with the restriction that it must be real for real ϕ . It can for instance be set equal to zero by definition. This freedom of choice corresponds to the possibility of adding a homogeneous solution: the dependence of $\langle r_\sigma | F_M^{so} \gamma \rangle$ on ϕ is exactly of the form stated in Theorem 4.1 .

It is easy to verify that r_σ as defined by (4.49) indeed solves (3.9), is real, and is symmetric in X . For greater reassurance we wish to check its invariance, even though this follows from our construction. Take $\phi = M_{\mu\nu}\psi$, $\psi \epsilon \mathcal{S}^{N+1}$. Construct the corresponding $F_M^{\cdot}$. From (4.43) we obtain $F_M^o = 0$. Let $\bar{F}_M$, $\bar{F}_M^1$, etc. , be the forms occurring in the expansion (4.45) of ψ . Note $M_{\mu\nu}\bar{F}_M^o = 0$. We obtain

$$\phi'' = M_{\mu\nu}\psi - \sum_{M=1}^{N} C[\gamma\, C^{-1} M_{\mu\nu} \bar{F}_M^1]$$

$$= M_{\mu\nu}(\psi - \sum_M \bar{F}_M \gamma) + \sum_M \{M_{\mu\nu}(\bar{F}_M^1 \gamma) - C[\gamma C^{-1} M_{\mu\nu} \bar{F}_M^1]\} \quad .$$

$\psi - \sum_1^N \bar{F}_M \gamma$ has vanishing derivatives up to order N at the origin.

Because of $M_{\mu\nu}\mathcal{W}_M \in \mathcal{W}_M$ the term $M_{\mu\nu}(\psi - \ldots)$ is in $\mathcal{S}_o^N$ and vanishes on r_σ (see the end of the 3rd step). Since $\phi'' \in \mathcal{S}_o^N$ the second term from above:

$$T_2 = \sum_M \{M_{\mu\nu}(\bar{F}_M^1\gamma) - C\,[\gamma C^{-1} M_{\mu\nu}\bar{F}_M^1]\} \quad ,$$

is also in $\mathcal{S}_o^N$. $M_{\mu\nu}$ commutes with C and C^{-1} , hence

$$T_2 = \sum_M \{M_{\mu\nu}(\bar{F}_M^1\gamma) - M_{\mu\nu}C\,[\gamma C^{-1}\bar{F}_M^1] + C[(M_{\mu\nu}\gamma)\,(C^{-1}\bar{F}_M^1)]\}$$

$$= \sum_M M_{\mu\nu}\{\bar{F}_M^1\gamma - C\,[\gamma C^{-1}\bar{F}_M^1]\} + \sum_M C\{(M_{\mu\nu}\gamma)\,(C^{-1}\bar{F}_M^1)\} \quad .$$

Both curly brackets belong to $\mathcal{S}_o^N \cap C^\infty$, hence $\langle r_\sigma | T_2\rangle = 0$, again because of the results of step 3 . This completes the invariance proof.

By exhibiting the ϕ_D -dependence of F_M explicitly we can write the development (4.47) similarly to (4.30):

$$\phi(\Xi) = \sum_{|D|\leq N} \phi_D\, \bar{\gamma}^D(\Xi) + \phi''(\Xi) \quad , \tag{4.50}$$

with $\bar{\gamma}^D \in \mathcal{D}$, $\bar{\gamma}^D = \gamma^D$ in a neighbourhood of the origin. The definition of r_σ is then, for $\phi \in \mathcal{S}^N$,

$$\langle r_\sigma(x,X)|\phi(x,X)\rangle = \sum_{i=1}^{n} \langle I_{i\sigma}(x,X)|\zeta_i(x,X)\phi''(x,X)\rangle + \sum_{|D|\leq N} r_\sigma^D\,\phi_D \ , \tag{4.51}$$

where the constants $r_\sigma^D = \langle r_\sigma|\bar{\gamma}^D\rangle$ are determined as described above.

A disturbing feature of (4.51) is that the test function ϕ

appears in it in such an essential way. We should much prefer to express the distribution r_σ in terms of the distributions $I_{i\sigma}$, without explicit reference to test functions. Such an expression will now be derived.

Consider the auxiliary functions $\zeta_i(x,X)$ defined in (4.24). They are C^∞ everywhere except at $x=x_1=\ldots=x_n$. Choose $\bar{\rho}(\xi)$ in $\mathcal{D}$ such that $\bar{\rho} \equiv 1$ in a neighbourhood of $\xi=0$, and $\bar{\rho}(-\xi)=\bar{\rho}(\xi)$. Define

$$\rho(x,X) = 1 - \prod_{i=1}^{n} \bar{\rho}(x-x_i) \prod_{1\leq i<j\leq n} \bar{\rho}(x_i-x_j) \quad . \tag{4.52}$$

ρ is C^∞, vanishes in a neighbourhood of $x=\ldots=x_n$, and is $\equiv 1$ outside a larger neighbourhood of the same manifold. Define

$$\rho_\kappa(x,X) = \rho(\kappa x,\kappa X) \quad , \qquad 0 < \kappa < \infty \quad . \tag{4.53}$$

ρ_κ has the same properties as ρ, the two relevant neighbourhoods depending on κ. The function

$$\zeta_{i\kappa}(x,X) = \rho_\kappa(x,X)\, \zeta_i(x,X) \tag{4.54}$$

is C^∞. $\zeta_{i\kappa}$ and its derivatives converge, for $\kappa\to\infty$, to ζ_i and its derivatives respectively, uniformly on all closed sets not intersecting the critical manifold $x=x_1=\ldots=x_n$. Closer inspection shows that for $\phi(x,X) \in \mathcal{S}_o^N$ the sequence $\zeta_{i\kappa}\phi$ is in $\mathcal{S}_o^N$ and converges for $\kappa\to\infty$ to $\zeta_i\phi$ in the topology of $\mathcal{S}^N$. Hence (4.51) can be written

$$\langle r_\sigma|\phi\rangle = \sum_i \lim_{\kappa\to\infty} \langle I_{i\sigma}|\zeta_{i\kappa}[\phi - \sum_D \phi_D \gamma^D]\rangle + \sum_D r_\sigma^D \phi_D \quad .$$

But $\zeta_{i\kappa}\phi \in \mathcal{J}^N$ and $\zeta_{i\kappa}\bar{\gamma}^D \in \mathcal{J}^N$, so that the two terms in the square bracket give, for finite κ, separately existing contributions:

$$\langle r_\sigma | \phi \rangle = \lim_{\kappa\to\infty} \{ \sum_i \langle I_{i\sigma} | \zeta_{i\kappa}\phi \rangle - \sum_D \phi_D \left[\sum_i \langle I_{i\sigma} | \zeta_{i\kappa}\bar{\gamma}^D \rangle - r_\sigma^D \right\} \quad . \tag{4.55}$$

Defining

$$R_{\sigma\kappa}^D = \sum_{i=1}^n \langle I_{i\sigma}\zeta_{i\kappa} | \bar{\gamma}^D \rangle - r_\sigma^D \tag{4.56}$$

and remembering

$$\phi_D = D\, \phi(\Xi) \Big|_{\xi_i=0} = (-1)^{|D|} \int d\Xi\, \phi(\Xi)\, D \prod_i \delta^4(\xi_i) \tag{4.57}$$

we obtain

$$\boxed{r_\sigma(x,X) = \lim_{\kappa\to\infty} \{ \sum_{i=1}^n \zeta_{i\kappa}(x,X)\, I_{i\sigma}(x,X) - \sum_{|D|\leq N} R_{\sigma\kappa}^D\, D \prod_{i=1}^n \delta^4(x-x_i)\}} \quad , \tag{4.58}$$

our final result. In p-space this becomes

$$\tilde{r}_\sigma(p,P) = \lim_{\kappa\to\infty} \{ (2\pi)^{n+1}\, \tilde{\zeta}_{i\kappa}(p,P) * \tilde{I}_{i\sigma}(p,P) - \delta^4(p+\Sigma p_i) \sum_D R_{\sigma\kappa}^D\, \mathcal{P}_D(P) \} \quad , \tag{4.59}$$

where

$$\tilde{\zeta}_{i\kappa}(p,P) = (2\pi)^{-\frac{5}{2}(n+1)} \int dx\, dX\, \exp\{i(px+\Sigma p_j x_j)\}\, \zeta_{i\kappa}(x,X) \tag{4.60}$$

and

$$\mathcal{P}_D(P) = (2\pi)^{4-\frac{5}{2}(n+1)} \quad (-1)^{|D|} \, D \exp\{-i \, \Sigma p_j \xi_j\}\Big|_{\xi_j=0} \quad . \qquad (4.61)$$

$\mathcal{P}_D$ is a form of degree $|D|$. The $*$ denotes the convolution product

$$A(P)*B(P) = \int dQ \; A(P-Q) \; B(Q) \quad .$$

Without going into details we wish to mention that the constants $R^D_{\sigma\kappa}$ are intimately connected with the subtraction constants (renormalisation constants) of the canonical formalism. The indeterminacy of the F^{so}_M -part of $R^D_{\sigma\kappa}$ (see (4.49)) corresponds to the ambiguity of these subtraction constants. A detailed, explicit discussion of this connection has not yet been undertaken.

Finally we must make a remark on the special case of the 2-point function. There we must satisfy the additional condition (2.58) which has not yet been considered. The $\tilde{r}_o(p,q)$ given by (3.3) satisfies (2.58), so that the condition becomes in the higher orders $\sigma > 1$

$$\tilde{r}_\sigma(p,q) = \delta^4(p+q) \; (q^2 - m^2)^2 \; F_\sigma(q) \qquad (4.62)$$

with F_σ analytic in $q^2 < 4m^2$. In order to meet this new condition we amend the $n=0$ case of Theorem 4.2 as follows.

<u>Theorem 4.3</u>

In order that the equation

$$r_\sigma(x,y) - r_\sigma(y,x) = I_\sigma(x,y) \qquad (4.63)$$

has a solution satisfying conditions i), ii) of Theorem 2.1 and condition (4.62) it is necessary and sufficient that $I_\sigma(x,y)$ is a tempered distribution with the following properties.

a) supp $I_\sigma(x,y) \subset \{(x-y)^2 \geq 0\}$.

b) I_σ is antisymmetrical:

$$I_\sigma(x,y) + I_\sigma(y,x) = 0 \quad . \tag{4.64}$$

c) I_σ is real.

d) I_σ is invariant under $\mathcal{P}_+^\uparrow$.

f) The Fourier transform $\tilde{I}_\sigma(p,q) = \delta^4(p+q)\, \mathtt{I}_\sigma(q)$ vanishes in $q^2 < 4m^2$.

Conditions a) - d) are the same as in Theorem 4.2 , which fact implies their necessity. We use (2.27) and (2.53):

$$\tilde{r}_\sigma(p,q) = \delta^4(p+q)\, \mathtt{r}_\sigma(q) \quad , \quad \tilde{r}_\sigma(q,p) = \delta^4(p+q)\, \mathtt{r}_\sigma(-q) \quad . \tag{4.65}$$

Similar arguments as used for $\mathtt{r}$ after (2.56) can be applied to $\mathtt{r}_\sigma$ and imply that $\mathtt{r}_\sigma(q) = \mathtt{r}_\sigma(-q)$ in $q^2 < 4m^2$. This proves the necessity of condition f).

The sufficiency proof proceeds again by construction of an explicit solution. From the proof of Theorem 4.2 we know that the expression (4.59), specialised to $n=1$, gives a solution of (4.63) satisfying all the requirements except possibly (4.62) . Let

$$\tilde{r}'_\sigma(p,q) = \delta^4(p+q)\, \mathtt{r}'_\sigma(q)$$

be of the form (4.59). $r'_\sigma(x,y)$ has support in $(x-y) \in \bar{V}_+$, so that $\mathtt{r}'_\sigma(q)$ is analytic in $(\mathrm{Im}\, q) \in V_-$, $\mathtt{r}'_\sigma(-q)$ in $(\mathrm{Im}\, q) \in V_+$. Condition f) and (4.63) yield

$$\tilde{r}'_\sigma(q) = \tilde{r}'_\sigma(-q) \qquad \text{for} \quad q^2 < 4m^2 \quad ,$$

so that, by the edge-of-the-wedge theorem, $\tilde{r}'_\sigma(q)$ is analytic for real q with $q^2 < 4m^2$. Because of its Lorentz invariance the analytic function $\tilde{r}'_\sigma$ depends only on q^2, so that we can expand in a neighbourhood of $q^2=m^2$:

$$\tilde{r}'_\sigma(q) = A + B\,(q^2-m^2) + (q^2-m^2)^2\, F_\sigma(q^2) \quad . \tag{4.66}$$

But the polynomial $A + B(q^2-m^2)$ is a solution of the homogeneous equation

$$\tilde{h}(q) - \tilde{h}(-q) = 0$$

with all the necessary subsidiary properties, hence

$$\tilde{r}_\sigma(q\ ) = (q^2-m^2)^2\, F_\sigma(q^2) \tag{4.67}$$

gives the desired solution $\tilde{r}_\sigma$ of (4.63) satisfying (4.62) .

V. SMALL DISTANCE BEHAVIOUR

We come back to the question of the ambiguities in the solution of the equation (3.9) for r_σ . We have stated in the discussion of Theorem 4.1 that the choice of r_1 fixes the interaction. This statement does not seem to make much sense at the moment, since ambiguities of the form (4.3) occur in any order σ . In order to get rid of them, at least to some extent, we must introduce an additional condition for r_σ . For this we choose a natural-looking assumption about the small distance behaviour of r_σ *.

Before we can formulate this condition we must make a short mathematical digression. Consider the space $\mathcal{S}(u_1,\ldots,u_\ell)$ of tempered test functions in ℓ variables. The linear mapping

$$\phi(U) \longrightarrow \phi_\lambda(U) = \phi(\lambda U) \quad , \qquad 0 < \lambda < \infty \, , \tag{5.1}$$

of $\mathcal{S}$ onto $\mathcal{S}$ is continuous in the topology of $\mathcal{S}$. We define a dual mapping $T(U) \longrightarrow T_\lambda(U)$, $T \in \mathcal{S}'(U)$, by

$$\langle T_\lambda | \phi \rangle = \lambda^\ell \langle T | \phi_\lambda \rangle \quad . \tag{5.2}$$

This mapping of $\mathcal{S}'$ onto $\mathcal{S}'$ is again linear and continuous. The family T_λ of distributions generated by a given $T \in \mathcal{S}'$ depends continuously on λ in $0 < \lambda < \infty$.

(5.2) reads in the improper but convenient integral notation

$$\int dU \, T_\lambda(U) \, \phi(U) = \lambda^\ell \int dU \, T(U) \, \phi_\lambda(U) = \int dV \, T(\lambda^{-1} V) \, \phi(V) \quad . \tag{5.3}$$

where we have substituted $\lambda u_i = v_i$. We can write symbolically

$$T_\lambda(u_1,\ldots,u_\ell) = T\left(\frac{u_1}{\lambda}, \ldots, \frac{u_\ell}{\lambda}\right) \, . \tag{5.4}$$

---

* Instead, we could impose restrictions on the high energy behaviour of $\tilde{r}_\sigma$. We prefer the x-space condition because the construction carried out in Chapter IV is based on x-space considerations.

Obviously, the limit $\lambda \to \infty$ of T has something to do with the behaviour of T for small u_i. (5.4) becomes under Fourier transformation

$$\tilde{T}_\lambda(w_1,\ldots,w_\ell) = \lambda^\ell \, \tilde{T}(\lambda w_1,\ldots,\lambda w_\ell) \quad . \tag{5.5}$$

This formula seems to show that the limit $\lambda \to \infty$ of T_λ is determined by the behaviour of $\tilde{T}$ for large values of its arguments. This appeareance is somewhat deceptive: The $\lambda\to\infty$ behaviour of $\tilde{T} = D\delta^\ell(W)$ depends very much on the derivative D, even though all these $\tilde{T}$ vanish for large w_i.

We prove

Lemma 5.1

For any $T(u_1,..,u_\ell)$ there exists a real constant $d > -\infty$ such that

$$\lim_{\lambda\to\infty} \lambda^\beta \, T_\lambda = 0 \tag{5.6}$$

for $\beta < d$ but not for $\beta > d$. The limit in (5.6) is to be understood in the sense of $\mathcal{S}'$. The value $d = +\infty$ is possible.

Proof. It is obvious that (5.6) holds for β if it holds for a β' with $\beta' > \beta$. Thus the condition (5.6) defines a cut in the real numbers. According to a theorem by Schwartz (Ref. [19], p.239) we can write T as a finite derivative of a continuous function of slow increase at infinity: there exist a differentation D in U of finite order $|D|$, a non-negative integer N, and a positive constant C, such that

$$T(U) = D\, f(U) \quad , \tag{5.7}$$

where f is a continuous function with

$$|f(U)| \leqslant C\,[1 + |U|^N] \quad . \tag{5.8}$$

Hence, for $\phi \in \mathcal{S}$:

$$\langle T|\phi\rangle = (-1)^{|D|} \int dU \frac{f(U)}{1+|U|^{N+\ell+1}} [1+|U|^{N+\ell+1}]\, D\phi(U) \quad ,$$

$$|\langle T \mid \phi\rangle| \leqslant \|\phi\|_{D,N+\ell+1} \cdot J \tag{5.9}$$

with

$$J = \int dU \; [1+|U|^{N+\ell+1}]^{-1} \; |f(U)| < \infty$$

and

$$\|\phi\|_{D,M} = \sup_U \mid [1+|U|^M]\, D\phi \mid \quad . \tag{5.10}$$

For ϕ_λ we have

$$\|\phi_\lambda\|_{D,M} = \text{Sup} \mid [1+|U|^M]\, D\phi(\lambda U)|$$

$$\leqslant \text{Sup}\, |D\phi(\lambda U)| + \text{Sup}|\,|U|^M\, D\phi(\lambda U)|$$

$$\leqslant \lambda^{|D|} \text{Sup}\, |D\phi(U)| + \lambda^{|D|-M}\, \text{Sup}|\,|U|^M\, D\phi(U)| \quad , \tag{5.11}$$

hence

$$|\langle T_\lambda|\phi\rangle| = \lambda^{\ell}|\langle T|\phi_\lambda\rangle| \leq J\lambda^{\ell}\{\lambda^{|D|}\|\phi\|_{D,0} + \lambda^{|D|-M}\|\phi\|_{D,M}\} \quad , \quad (5.12)$$

i.e. $\lambda^{\beta}\langle T_\lambda|\phi\rangle \to 0$ for $\beta < -\ell-|D|$. This limit is attained uniformly on bounded sets of $\mathcal{S}$, a bounded set being defined by the requirement that all $\|\phi\|_{D,M}$ are bounded by some given finite values (Ref. [19], p.235). This proves the lemma and shows that $d > -\ell-|D| > -\infty$.

The value d defined by Lemma 5.1 is called the <u>scaling degree</u>*, or <u>s-degree</u> for short, of the distribution T . We abbreviate it as SD(T) .

<u>Examples</u>:

1) A form

$$T(U) = \sum_{\Sigma\alpha_i = A} c_{\alpha_i} u_i^{\alpha_i}$$

of degree A has s-degree A . This result is not changed if T is multiplied by a logarithmic factor ($\log|u_i|$ or similar), i.e. logarithmic singularities are counted the same as bounded functions in this definition of the s-degree. This is important for the perturbative applications.

2) A derivative

$$D \prod_{i=1}^{\ell} \delta(u_i)$$

of a δ-function, D of order $|D|$, has s-degree $-\ell-|D|$.

3) Any distribution whose support does not contain the origin has

* We prefer this name to the appellation "dimension" that has recently become popular for this quantity in a different context. We feel that "dimension" should be allowed to retain its time-honoured meaning fixing the place of T within a specified system of units.

s-degree ∞ .

4) Let T_1 , $T_2 \in \mathcal{S}'$ with s-degrees $d_{1,2}$. If $d_1 < d_2$ then the s-degree d of $T_1 + T_2$ is $d=d_1$. If $d_1=d_2$ then $d \geq d_1$, since the leading terms in T_1 and T_2 may cancel.

We can now give the additional condition for r_σ promised above: <u>we demand that the solution r_σ of (3.9), for $\sigma > 2$, be chosen of the maximal possible s-degree</u>. In view of the above examples this means that we ask for as smooth a small distance behaviour as is possible. This condition looks natural and is often used in similar forms also outside perturbation theory.

The simplest application of this principle in our context is the following: if $I_\sigma(x_1,..,x_n) = 0$ (and this is so in any fixed order σ for all n except finitely many), then $r_\sigma = 0$ is to be chosen as solution. What the maximal dimension is if $I_\sigma \neq 0$ we learn from an expanded version of Theorem 4.2 :

<u>Theorem 5.2</u>

Let $I_\sigma(x,y,x_1,..,x_n)$ be of s-degree d and satisfy conditions a) - e) of Theorem 4.2 . Then there exists a solution r_σ of (3.9) with s-degree d satisfying conditions i) and ii) of Theorem 2.1 . No solution r_σ with s-degree $>d$ exists.

The second part of the theorem is obvious in view of the fourth example given above. From $SD(r_\sigma) > d$ we deduce $SD(I_\sigma)>d$ contrary to assumption.

The proof of the existence of a solution with s-degree d is more complicated. We use the ideas and notations of Chapter IV .

Note that if the s-degree of the distribution $I_\sigma(x,x_1,..,x_n)$ is d , then the s-degree of its expression $I_\sigma(\xi_1,..,\xi_n)$ in terms of the differences $\xi_i=x-x_i$ is also d . This follows from (5.4)

Let $\phi(\Xi) \in \mathcal{S}_o^N$. Then $\phi_\lambda(\Xi) \in \mathcal{S}_o^N$, and $\langle r_\sigma|\phi_\lambda\rangle$ is given by (4.22):

$$\langle r_\sigma \phi_\lambda\rangle = \sum_i \langle I_{i\sigma}|\zeta_i\phi_\lambda\rangle \quad . \tag{5.13}$$

But ζ_i is scale invariant, hence $\zeta_i \phi_\lambda = (\zeta_i \phi)_\lambda$, and

$$\lambda^{\beta+4n} \langle r_\sigma | \phi_\lambda \rangle = \sum_i \lambda^{\beta+4n} \langle I_{i\sigma} | (\zeta_i \phi)_\lambda \rangle \to 0 \quad \text{for } \lambda \to \infty , \ \beta < d, \tag{5.14}$$

because of the assumed $SD(I_\sigma)$.

Consider now $\phi \in \mathcal{S}^N$, which implies $\phi_\lambda \in \mathcal{S}^N$. Obviously

$$(\phi_\lambda)_D = \lambda^{|D|} \phi_D \tag{5.15}$$

and

$$\gamma_\lambda^D = \lambda^{|D|} \gamma^D \tag{5.16}$$

in a neighbourhood of $\xi_i = 0$, so that $\lambda^{-|D|} \gamma_\lambda^D(\Xi)$ has the same general properties as γ^D itself. According to Chapter IV we can write

$$\langle r_\sigma | \phi_\lambda \rangle = \sum_i \langle \zeta_i I_{i\sigma} | \phi_\lambda - \sum_D \phi_D \gamma_\lambda^D \rangle + \sum_D \phi_D \langle r_\sigma | \gamma_\lambda^D \rangle . \tag{5.17}$$

The first term is of the form (5.13), hence has the correct asymptotic behaviour for $\lambda \to \infty$. The second term has the correct behaviour wherever it is uniquely defined: on $C[\gamma C^{-1} F_M^1]$ it vanishes, on $F_M^{ao} \gamma$ r_σ is equal to I_σ , in both cases we have s-degrees $\geq d$. We must show that the undetermined parts of r_σ can be chosen such that r_σ has s-degree d . This means that there is a solution r_σ such that

$$\lim_{\lambda \to \infty} \lambda^{\beta+4n} \langle r_\sigma | \gamma_\lambda^D \rangle = 0 \quad \text{for } \beta < d . \tag{5.18}$$

Let r'_σ be a solution as constructed in Chapter IV, not necessarily with the desired s-degree. On $\mathcal{J}_0^N$ we have $r_\sigma = r'_\sigma$.

Choose two constants τ' , τ'' , with $0<\tau'<\tau''<\infty$. The functions $\tau^{-|D|}\gamma_\tau^D - \gamma^D$, D fixed, τ varying in $[\tau',\tau'']$, form a bounded set in $\mathcal{J}_0^N$. For $\beta < d$ we can find a constant $c_\beta > 0$ such that

$$\lambda^{\beta+4n} \, | \langle r'_\sigma | \, \tau^{-|D|}\gamma^D_{\tau\lambda} - \gamma^D_\lambda \rangle \, | < c_\beta \quad \text{for} \quad 1<\lambda<\infty \; . \tag{5.19}$$

Define

$$R_\lambda = \lambda^{-|D|} \, \langle r'_\sigma | \gamma^D_\lambda \rangle \quad , \tag{5.20}$$

so that (5.19) becomes

$$|R_{\tau\lambda} - R_\lambda| < c_\beta \, \lambda^{-\beta'} \qquad , \beta' = \beta+4n+|D| \; . \tag{5.21}$$

Choose τ' , τ'' , such that $1<\tau'<\tau'^2<\tau''<\infty$, and the positive constant $\mu>1$ such that $1 - \frac{1}{\mu} \geq \tau'^{-\beta'}$. We wish to prove that a real constant a exists with

$$|R_\lambda - a| \leq c_\beta \, \mu \, \lambda^{-\beta'} \tag{5.22}$$

for all $\lambda>1$.

For b sufficiently small we can find a $\lambda_1 > 1$ with

$$R_{\lambda_1} > b + c_\beta \, \mu \, \lambda_1^{-\beta'} \tag{5.23}$$

and for b sufficiently large a $\lambda_2 > 1$ with

$$R_{\lambda_2} < b - c_\beta \mu \lambda_2^{-\beta'} \quad . \tag{5.24}$$

The non-existence of an a satisfying (5.22) is equivalent to the existence of a triple b, λ_1, λ_2 such that both inequalities (5.23) and (5.24) are satisfied. (5.23) can be fulfilled for b in an open left semi-axis, (5.24) for b in an open right semi-axis. If these two semi-axes do not overlap, then the a lying in between satisfy (5.22). We wish to show that in fact the semi-axes do not overlap. Assume the contrary: that the triple b, λ_1, λ_2 exists. For $\tau'<\tau<\tau''$ we estimate with the help of (5.21) and (5.23)

$$R_{\tau\lambda_1} = R_{\tau\lambda_1} - R_{\lambda_1} + R_{\lambda_1} \geq - c_\beta \lambda_1^{-\beta'} + b + c_\beta \mu \lambda_1^{-\beta'}$$

$$\geq b + c_\beta \mu (\tau\lambda_1)^{-\beta'} \quad :$$

$R_{\tau\lambda_1}$ satisfies again (5.23). Iteration of this procedure tells us that

$$R_\lambda \geq b + c_\beta \mu \lambda^{-\beta'} \tag{5.25}$$

whenever λ/λ_1 can be written as a product of factors τ_i lying in $[\tau',\tau'']$. Under our assumption $\tau''>\tau'^2$ such a factorisation exists for all $\lambda>\tau'\lambda_1$.

In the same way we obtain from (5.24)

$$R_{\tau\lambda_2} = R_{\tau\lambda_2} - R_{\lambda_2} + R_{\lambda_2} \leq c_\beta \lambda_2^{-\beta'} + b - c_\beta \mu (\tau\lambda_2)^{-\beta'}$$

for $\tau'<\tau<\tau''$ and by iteration

$$R_\lambda \leq b - c_\beta \mu \lambda^{-\beta'} \tag{5.26}$$

for $\lambda > \tau' \lambda_2$. Combining (5.25) and (5.26) we see that for λ sufficiently large

$$b + c_\beta \, \mu \lambda^{-\beta'} \leq R_\lambda \leq b - c_\beta \, \mu \lambda^{-\beta'} ,$$

a contradiction. This proves the existence of an a satisfying (5.22). It is clear that this a is uniquely fixed and does not depend on β if $\beta > 0$. This is possible for $d + 4n + |D| > 0$. In this case we have

$$\lambda^{4n+\beta} \left| \langle r'_\sigma | \gamma^D_\lambda \rangle - a \lambda^{|D|} \right| \leq c_\beta \, \mu .$$

We define a distribution

$$h^D_\sigma = (-1)^{|D|} \, a \quad D \prod_i \delta^4(\xi_i) \tag{5.27}$$

and obtain

$$\lambda^{4n+\beta} \left| \langle r'_\sigma - h^D_\sigma | \gamma^D_\lambda \rangle \right| < c_\beta \, \mu . \tag{5.28}$$

(Note that a depends on D and σ even though this dependence has not been expressed explicitly.) We have remarked above that the asymptotic behaviour of $\hat{r}'_\sigma$ is the desired one where r_σ is defined uniquely. It follows from this that

$$h^{|D|}_\sigma = \sum_{\substack{D \\ D=\mathrm{const}}} h^D_\sigma \tag{5.29}$$

is a solution of the homogeneous equation (4.1) satisfying all subsidiary conditions. Replacement of $\langle r'_\sigma | \gamma^D \rangle$ by

$$\langle r_\sigma | \gamma^D \rangle = \langle r'_\sigma - h^D_\sigma | \gamma^D \rangle \tag{5.30}$$

will therefore give a solution with the desired property (5.18). If $d + 4n + |D| \leq 0$, then <u>any</u> solution r_σ , in particular the initially chosen r'_α , will satisfy (5.18): from

$$|R_\lambda - a| \leq c_\beta \, \mu \, \lambda^{-\beta'}$$

we obtain

$$|\langle r_\sigma | \gamma^D_\lambda \rangle| \leq a \lambda^{|D|} + c_\beta \, \mu \, \lambda^{-\beta - 4n} \quad ,$$

whence (5.18) follows since $\beta' < 0$ means $\beta + 4n < -|D|$.

With these results we can now give the solution r_σ of dimension d whose existence is claimed in Theorem 5.2 . Define, for $\phi \in \mathcal{S}^N$,

$$\phi'_\lambda = \phi - \sum_{\substack{D \\ -d-4n < |D| \leq N}} \phi_D \, \lambda^{-|D|} \gamma^D_\lambda - \sum_{\substack{D \\ |D| \leq -d-4n}} \phi_D \, \gamma^D \tag{5.31}$$

and

$$\langle r_\sigma | \phi \rangle = \langle r_\sigma | \phi'_\lambda \rangle + \sum_{\substack{D \\ -d-4n < |D| \leq N}} \phi_D \, \lambda^{-|D|} \langle r_\sigma | \gamma^D_\lambda \rangle$$

$$+ \sum_{\substack{D \\ |D| \leq -d-4n}} r^D_\sigma \, \phi_D \quad . \tag{5.32}$$

The first term is defined because of $\phi'_\lambda \in \mathcal{S}^N_o$. The r^D_σ are the same as in (4.51). The $|D| > -d-4n$ terms defined by (5.30) vanish in the limit $\lambda\to\infty$. The full right-hand side is independent of λ, so that

$$\langle r_\sigma|\phi\rangle = \lim_{\lambda\to\infty} \langle r_\sigma|\phi'_\lambda\rangle + \sum_{|D|\leq -d-4n} r^D_\sigma \phi_D \,, \tag{5.33}$$

and the limit exists. The explicit dependence on ϕ can be eliminated from this expression in the same way as it was from (4.51). With the definition

$$R^D_{\sigma\kappa\lambda} = \lambda^{-|D|} \sum_i \langle I_{i\sigma}\zeta_{i\kappa} \mid \gamma^D_\lambda\rangle \tag{5.34}$$

we obtain the final result, in analogy to (4.58),

$$\boxed{r_\sigma(x,X) = \lim_{\lambda\to\infty} \lim_{\kappa\to\infty} \Big\{ \sum_i \zeta_{i\kappa}(x,X)\, I_{i\sigma}(x,X) - \sum_{\substack{D\\ -d-4n<|D|\leq N}} R^D_{\sigma\kappa\lambda}\, D \prod_{j=1} \delta^4(x-x_j) - \sum_{\substack{D\\ |D|\leq -d-4n}} R^D_{\sigma\kappa}\, D \prod_j \delta^4(x-x_j) \Big\} .} \tag{5.35}$$

This is the solution singled out by our maximality requirement for $SD(r_\sigma)$. If $d\leq -4n$ this solution is still not unique but contains the ambiguities inherent in the choice of the r^D_σ, $|D|\leq -4n-d$. This is equivalent to the possibility of adding homogeneous solutions (4.3) with $|D|\leq -d-4n$. These homogeneous terms are of dimension $-|D|-4n\geq d$, hence their addition will not decrease the s-degree of r_σ.. This ambiguity is not specific to our approach but manifests itself also in the canonical formalism as ambiguity in the coefficients of the counter terms that are necessary to cancel infinities. We shall see in Chapter VIII that for the so-called renormalisable theories we do indeed not get any more ambiguities than in the canonical way of doing things.

For the 2-point function the optimal solution (5.35) does in general not satisfy condition (4.62), at least if $d > -6$. For $d \leq -6$ we can always satisfy (4.62) as in Theorem 4.3 by adding a polynomial of second degree to $\tilde{r}_\sigma(q)$, without decreasing $SD(r_\sigma)$. For $d > -6$ such an addition must also be performed in general, but does then decrease $SD(r_\sigma)$ to -6 (or possibly -4). Thus, for $d > -6$, the s-degree of the solution $r_\sigma(x,y)$ satisfying the small-distance smoothness condition is in general -6, not d. Nevertheless, this solution is unique. Ambiguities occur only for $d<-6$. They are in $\tilde{r}_\sigma(q)$ of the form

$$\sum_{2\leq k\leq -\frac{d+4}{2}} c_k(q^2-m^2)^k \quad . \tag{5.36}$$

VI. THE PROPERTIES OF I_σ

The recursion program outlined in Chapter III is implemented by the contents of Chapters IV and V, provided that I_σ calculated from (3.10) and (3.11) fulfills the assumptions of Theorems 4.2 and 4.3. To show this is the problem which will occupy us now. For the moment we assume the existence of the integrals (3.11) and discuss their relevant properties. Existence will then be proved in the following chapter.

Let $r_\tau(x,x_1,\ldots,x_n)$, $\tau < \sigma$, be known and satisfy the conditions of Theorem 2.3 in their perturbative form. Calculate $J_{\ell L\sigma}$, I_σ, from (3.11) and (3.10). We must show that these I_σ satisfy conditions a)-e) of Theorems 4.2 and 4.3 respectively.

It is obvious that conditions d), e) and (4.9) are satisfied. The reality condition c) follows easily from $[\Delta_+(\xi)]^* = - \Delta_+(-\xi)$ which implies

$$J^*_{\ell L\sigma}(x,y,X) = (-1)^\ell J_{\ell R\sigma}(y,x,X) \quad . \tag{6.1}$$

From among the less trivial points we attack first the Jacobi-like identities (4.10) . For simplicity of notation we consider only the special case of the 3-point function. The proof for the general case can be derived from this simply by inserting x_i's in the appropriate places.

Define

$$[xz,y] = - i \sum_{\ell=1}^{\infty} \frac{i^\ell}{\ell!} J_{\ell z\sigma}(x,y,z) \quad ,$$

$$[x,yz] = -i \sum_{\ell=1}^{\infty} \frac{i^\ell}{\ell!} J_{\ell\,\text{[illegible]}\,\sigma}(x,y,z) \quad . \tag{6.2}$$

In words: $[xz,y]$ contains the $J_{..}$ in which the variable z occurs in the first r-factor (together with x), $[x,yz]$ those with z in the same r-factor as y . Then

$$I_\sigma(x,y,z) = [xz,y] + [x,yz] - [yz,x] - [y,xz] \quad . \tag{6.3}$$

We wish to prove that

$$\begin{aligned} L &= I_\sigma(x,y,z) + I_\sigma(y,z,x) + I_\sigma(z,x,y) \\ &= ([xz,y] - [zx,y]) + ([x,yz] - [x,zy]) + ([zy,x] - [yz,x]) \\ &\quad + ([y,zx] - [y,xz]) + ([yx,z] - [xy,z]) + ([z,xy] - [z,yx]) \end{aligned} \tag{6.4}$$

vanishes. The round brackets have been introduced to show that the terms of L can be combined into pairs which are all of a similar structure. Take such a pair, e.g.

$$[xz,y] - [zx,y] = -i \sum_{\ell} \frac{i^{\ell}}{\ell!} \sum_{\tau=1}^{\sigma-1} \int \prod_{i=1}^{\ell} \{du_i \, dv_i \, \Delta_+(u_i - v_i)\}$$

$$\times \quad \{r_\tau(x,z,U) - r_\tau(z,x,U)\} \, r_{\sigma-\tau}(y,V) \quad .$$

The equation (3.9) with the index σ replaced by τ is assumed to be valid for $\tau<\sigma$, hence the second curly bracket in the above expression is equal to $I_\tau(x,z,U)$. Substitution of the explicit expression for I_τ yields

$$[xz,y] - [zx,y] = [[x,z],y] - [[z,x],y] \tag{6.5}$$

with

$$[[x,z],y] = - \sum_{\tau+\rho=\sigma} \sum_{\ell=1}^{\infty} \frac{i^{\ell}}{\ell!} \sum_{\tau'+\tau''=\tau} \sum_{\alpha,\beta} \sum_{k=1}^{\infty} \frac{i^{k}}{k!}$$

$$\times \int dU \, dV \, dU' \, dV' \, r_{\tau'}(x,U_\alpha,U') \prod_{1}^{k} \Delta_+(u'_j - v'_j) \quad r_{\tau''}(z,U_\beta,V')$$

$$\times \prod_{1}^{\ell} \Delta_+(u_i - v_i) \, r_\rho(y,V) \quad . \tag{6.6}$$

U and V contain ℓ variables, U' and V' k variables. The (α,β)-sum extends over all partitions of U into two complementary subsets U_α and U_β . The ℓ- and k-sums contain only finitely many non-vanishing terms, so that reorderings of the sum are allowed.

In the same way we obtain

$$[x,yz] - [x,zy] = [x,[y,z]] - [x,[z,y]] \tag{6.7}$$

with

$$[x,[y,z]]$$

$$= - \sum_{\tau+\rho=\sigma} \sum_{\ell=1}^{\infty} \frac{i^{\ell}}{\ell!} \sum_{\rho'+\rho''=\rho} \sum_{\alpha,\beta} \sum_{k=1}^{\infty} \frac{i^k}{k!} \int dU\, dV\, dU'\, dV' \quad r_\tau(x,U)$$

$$\times \prod_1^{\ell} \Delta_+(u_i-v_i)\; r_{\rho'}(y,V_\alpha,U')\; \prod_1^{k} \Delta_+(u_j'-v_j')\; r_{\rho''}(z,V_\beta,V') \qquad (6.8)$$

with analogous notation to (6.6)

Because of the symmetry of $r_\rho(y,V)$ in V we see that all terms in the (α,β)-sum in (6.6) with the same number of elements in U_α give the same contribution. We can therefore replace $\Sigma_{\alpha,\beta}$ by $\sum_{\ell'=0}^{\ell} \binom{\ell}{\ell'}$ with $U_\alpha = (u_1,\ldots,u_{\ell'})$, $U_\beta = (u_{\ell'+1},\ldots,u_\ell)$. We rename the variables as follows:

$$u_1,\ldots,u_{\ell'} \longrightarrow u_1,\ldots,u_{\ell'} \;,$$

$$v_1,\ldots,v_{\ell'} \longrightarrow u_1',\ldots,u_{\ell'}' \;,$$

$$u_{\ell'+1},\ldots,u_\ell \longrightarrow v_1,\ldots,v_{\ell''} \;, \qquad \ell'' = \ell - \ell' \;,$$

$$v_{\ell'+1},\ldots,v_\ell \longrightarrow v_1',\ldots,v_{\ell''}' \;,$$

$$u_1',\ldots,u_k' \longrightarrow w_1,\ldots,w_k \;,$$

$$v_1',\ldots,v_k' \longrightarrow w_1',\ldots,w_k' \;,$$

and reorder the sums in (6.6) to

$$[[x,z],y]$$

$$= - \sum_{\tau+\rho+\omega=\sigma} \sum_{\ell'=1}^{\infty} \sum_{\ell''=1}^{\infty} \sum_{k=1}^{\infty} \frac{i^{\ell'+\ell''+k}}{\ell'!\,\ell''!\,k!} \int dU\, dU'\, dV\, dV'\, dW\, dW'$$

$$\times\; r_\tau(x,U,W)\; r_\rho(z,V,W')\; r_\omega(y,U',V')\; \prod_1^{\ell'} \Delta_+(u_i-u_i')\; \prod_1^{\ell''} \Delta_+(v_i-v_i')\; \prod_1^{k} \Delta_+(w_i-w_i') \;.$$

An analogous procedure applied to $[x,[z,y]]$ gives the same result:

$$[[x,z],y] = [x,[z,y]]$$

whence $L = 0$ follows at once.

Next we dispose of condition f) of Theorem 4.3 . (3.11) becomes in p-space for $n=0$

$$\tilde{\mathcal{J}}_{\ell L\sigma}(p,q) = \sum_{\tau=1}^{\sigma-1} \int \prod_{1}^{\ell} \{dk_i\ \delta_+(k_i)\}\ \tilde{r}_\tau(p,-K)\ \tilde{r}_{\sigma-\tau}(q,K) \quad . \tag{6.9}$$

Because of the δ^4-factor in $\tilde{r}_{\sigma-\tau}$ the integrand is $\neq 0$ only for $q = -\Sigma k_i$. But k_i must lie on the positive mass shell $k_i^2 = m^2$, $k_{io}>0$, because of the δ_+ factors. Hence $\tilde{\mathcal{J}}_{\ell L\sigma} = 0$ unless $q^2 \geq (\ell m)^2$. But $\tilde{\mathcal{J}}_{1L\sigma} = 0$ because of (3.3) and (4.62), the lowest non-vanishing term is the $\ell = 2$ one, and $\tilde{I}_\sigma(p,q) = 0$ for $q^2<4m^2$, q.e.d.

This leaves the support condition a) for $I_\sigma(x,y,x_1,..,x_n)$ to be proved. The condition: $(x-x_i)\in\overline{V}_+$ or $(y-x_i)\in\overline{V}_+$ for all i , follows at once from (3.11) and the support of r_τ . It remains to be seen that $I_\sigma = 0$ for $(x-y)^2< 0$.

From (4.9) and (4.10) we obtain

$$I_\sigma(x,y,z,\ldots) = I_\sigma(z,y,x,\ldots) + I_\sigma(x,z,y,\ldots) \quad . \tag{6.10}$$

Assume $(x-y)^2 < 0$. The left-hand side of (6.10) is different from zero only if $(x-z)\in\overline{V}_+$ or $(y-z)\in\overline{V}_+$, according to the part of the support condition which is already proved. To fix the ideas, assume $(y-z)\in\overline{V}_+$ (the other case is treated in exactly the same way) .

We have then $(z-x)\notin\overline{V}_+$, since from $(z-x)\in\overline{V}_+$ and $(y-z)\in\overline{V}_+$ it would follow that $(y-x)\in\overline{V}_+$ which is excluded by assumption. Hence $I_\sigma(z,y,x,..) = 0$ because neither $(z-x)\in\overline{V}_+$ nor $(y-x)\in\overline{V}_+$. In the same way: $I_\sigma(x,z,y,..) = 0$ because neither $(x-y)\in\overline{V}_+$ nor $(z-y)\in\overline{V}_+$, <u>except if $y=z$</u> . In the $(x-z)\in\overline{V}_+$ case we get the exceptional points $x=z$.

We learn from this that $I_\sigma(x,y,z,..) = 0$ for $(x-y)^2 < 0$, except possibly in the points $x=z$ or $y=z$. But we could have chosen any x_i in $I_\sigma(x,y,x_1,..,x_n)$ to be our z , so that we obtain: $I_\sigma(x,y,X) = 0$ for $(x-y)^2 < 0$, <u>except possibly in the points where each x_i coincides with either x or y</u> . This possibility must yet be excluded, a little problem that turns out to be amazingly obstinate.

Let X_L and X_R be two complementary subsets of $X = \{x_1,..,x_n\}$. Consider the points where all $x_i\in X_L$ coincide with x , all $x_i\in X_R$ with y , $(x-y)^2 < 0$. $I_\sigma(x,y,X)$ may possibly be non-zero there. Because of the support of r_τ , $\tau<\sigma$, non-vanishing contributions to I_σ can, however, only come from

$$I_{L\sigma}(x,y,X) = -i \sum_{\ell=1}^{\infty} \frac{i^\ell}{\ell!} \left[J_{\ell L\sigma}(x,y,X) - J_{\ell R\sigma}(y,x,X)\right] \quad .(6.11)$$

Let $\mathcal{S}_s(x,y,X)$ be the linear subspace of $\mathcal{S}(x,y,X)$ consisting of the test functions $\phi(x,y,X)$ with support in $(x-y)^2 \leqslant 0$. We want to prove that $I_{L\sigma}$ vanishes on $\mathcal{S}_s$. Call $S(x,y,X)$ the restriction of $I_{L\sigma}(x,y,X)$ to $\mathcal{S}_s$. It is a continuous linear form on $\mathcal{S}_s$, i.e. it is an element of the dual space $\mathcal{S}_s'$ of $\mathcal{S}_s$. Most of the well-known theorems about $\mathcal{S}'$ are also valid in $\mathcal{S}_s'$, with possibly some obvious alterations. Proofs can usually be taken over literally from the $\mathcal{S}'$-case. In particular, the theorem on the structure of distributions with support on a subspace (Ref. [19],p.101, Theorème XXXVI) remains valid. Applied to our problem this theorem states that S is of the form

$$S(x,y,X) = \sum_D T_D(x-y)\, D \prod_{X_L} \delta^4(x-x_i) \prod_{X_R} \delta^4(y-x_i) \quad . \tag{6.12}$$

Here the D are differentations in x_i of order $|D|$, $T(x-y) \in \mathcal{S}'_s(x,y)$, and the sum extends over a finite number of terms only. With $\partial_{i\mu} = \partial/\partial x_i^\mu$ we can write (6.12) more explicitly

$$S(x,y,X) = \sum_\alpha T_\alpha^{\mu_1^1 \ldots \mu_{\alpha_1}^1 \mu_1^2 \ldots \mu_{\alpha_2}^2 \ldots\ldots \mu_{\alpha_n}^n}(x-y)$$

$$\times\ \partial_{1\mu_1^1} \ldots \partial_{n\mu_{\alpha_n}^n} \prod_{X_L} \delta^4(x-x_i) \prod_{X_R} \delta^4(y-x_i) \tag{6.13}$$

α is the multi-index $\{\alpha_1,..,\alpha_n\}$. $T_\alpha^{\cdots\cdot}$ is symmetric in the indices with a fixed i :

$$T_\alpha^{\cdot\cdot\mu_\rho^i\cdot\cdot\mu_\sigma^i\cdot\cdot} = T_\alpha^{\cdot\cdot\mu_\sigma^i\cdot\cdot\mu_\rho^i\cdot\cdot} \quad .$$

S is invariant under the restricted Lorentz group. This implies that the $T_\alpha^{\cdots}(\xi)$, for fixed α, form a contravariant tensor. We prove that such a tensor distribution can be written as a covariant polynomial in ξ multiplied with an invariant distribution.

Lemma 6.1

Let $T^{\mu_1 \ldots \mu_N}(\xi) \in \mathcal{S}'_s$ transform under Lorentz transformations $\Lambda \in L_+^\uparrow$ as

$$T^{\mu_1 \ldots \mu_N}(\Lambda\xi) = \prod_{i=1}^{N} \Lambda^{\mu_i}{}_{\nu_i}\ T^{\nu_1 \ldots \nu_N}(\xi) \quad . \tag{6.14}$$

Then

$$T^{\mu_1 \ldots \mu_N}(\xi) = \xi^{\mu_1} \ldots \xi^{\mu_N} T(\xi) \tag{6.15}$$

with $T(\xi) \in \mathcal{S}'_s$ and

$$T(\Lambda\xi) = T(\xi) \quad . \tag{6.16}$$

We prove the lemma first in the case $N=1$. The infinitesimal form of (6.14) is in this case

$$M^{\mu}_{\nu}\, T^{\nu}(\xi) = T^{\mu}(\xi) \quad \text{, no summation over } \nu \text{ , } \mu \neq \nu \quad , \tag{6.17}$$

with

$$M^{\mu}_{\nu} = \xi^{\mu}\partial_{\nu} - \partial^{\mu}\xi_{\nu} \quad , \tag{6.18}$$

$$\xi_{\mu} = g_{\mu\nu}\xi^{\nu} \quad , \quad \partial^{\mu} = g^{\mu\nu}\partial_{\nu} \quad , \tag{6.19}$$

$g_{\mu\nu}$, $g^{\mu\nu}$ the Minkowski tensor. Note that M^{μ}_{ν} is also defined for equal indices, where it becomes

$$M^{\nu}_{\nu} = -1 \text{ , no summation} \quad , \tag{6.20}$$

so that

$$M^{\mu}_{\mu}\, T^{\mu}(\xi) = -T^{\mu}(\xi) \quad \text{, no summation} \quad . \tag{6.21}$$

From (6.17) and (6.21) we obtain

$$T^{\mu}(\xi) = \tfrac{1}{2}\, M^{\mu}_{\nu}\, T^{\nu}(\xi) \quad , \tag{6.22}$$

where the summation convention has been reinstated. Substituting (6.18) gives

$$T^{\mu}(\xi) = \tfrac{1}{2}\, \xi^{\mu}T_1(\xi) + \tfrac{1}{2}\, \partial^{\mu}T_2(\xi) \quad , \tag{6.23}$$

where

$$T_1(\xi) = \partial_\nu T^\nu(\xi) \quad ,$$
$$T_2(\xi) = - \xi_\nu T^\nu(\xi) \tag{6.24}$$

are invariant distributions.

This far the proof holds in $\mathcal{S}'$ as well as in $\mathcal{S}'_s$. From now on we use the fact that we are in $\mathcal{S}'_s$. Any invariant distribution $T(\xi) \in \mathcal{S}'_s$ can be written as distribution in the invariant ξ^2 by the following procedure [28,29]. Let $\phi(\xi) \in \mathcal{S}_s$. We define

$$\bar{\Phi}(\tau) = \int d^4\xi \; \delta(\xi^2 - \tau) \; \phi(\xi) \quad . \tag{6.25}$$

$\bar{\Phi}(\tau)$ is in $\mathcal{S}$ and has its support in $\tau \leq 0$. Call this space of test functions $\mathcal{S}_s(\tau)$. Let $\gamma(\xi)$ be a C^∞-function with $\bar{\gamma}(\tau) \equiv 1$. Such functions exist. We can define a distribution $\bar{T}(\tau)$ on $\mathcal{S}_s(\tau)$ by

$$\langle \bar{T}(\tau) \mid \phi(\tau) \rangle = \langle T(\xi) \mid \gamma(\xi) \;\; \bar{\Phi}(\xi^2) \rangle \quad . \tag{6.26}$$

$\gamma(\xi)\ \bar{\Phi}(\xi^2)$ is in $\mathcal{S}_s(\xi)$. With the definition (6.26) we have

$$\langle T(\xi) \mid \phi(\xi) \rangle = \langle \bar{T}(\tau) \mid \bar{\Phi}(\tau) \rangle \quad . \tag{6.27}$$

The bar over T in $\bar{T}$ will now be dropped, and we write in an improper but understandable way

$$T(\xi) \;=\; T(\xi^2) \quad . \tag{6.28}$$

From (6.25) we obtain

$$\overline{\partial_\mu\phi}(\tau) = \int d^4\xi\ \delta(\tau-\xi^2)\ \partial_\mu\phi(\xi)$$

$$= -\int d^4\xi\ \phi(\xi)\ \partial_\mu\delta(\tau-\xi^2)$$

$$= 2\int d^4\xi\ \delta'(\tau-\xi^2)\xi_\mu\ \phi(\xi) \quad ,$$

and this is in $\mathscr{S}_s(\tau)$ if $\phi \in \mathscr{S}_s(\xi)$. This implies, in our sloppy notation,

$$\partial_\mu T(\xi^2) = 2\ \xi_\mu T'(\xi^2) \quad , \tag{6.29}$$

where $T'(\tau) \in \mathscr{S}'_s$ is the derivate of $T(\tau) \in \mathscr{S}'_s$.

We apply this result to T_2 and obtain from (6.23)

$$T^\mu(\xi) = \xi^\mu T(\xi) \cdot \tag{6.30}$$

with

$$T(\xi) = \tfrac{1}{2}\ T_1(\xi) + T'_2(\xi^2) = T(\xi^2) \tag{6.31}$$

an invariant distribution in $\mathscr{S}'_s$. This proves the lemma for $N=1$.

The general case is proved by induction with respect to N .

Assume the lemma to be true for ranks $1,\dots,N-1$. Let $a^{(1)},\dots,a^{(N-1)}$, be arbitrary 4-vectors with covariant components $a_\mu^{(i)}$. Then

$$T^{\mu_N}(\xi) = a^{(1)}_{\mu_1}\dots a^{(N-1)}_{\mu_{N-1}}\, T^{\mu_1\dots\mu_N}(\xi)\cdot$$

is a contravariant vector, and hence can be written in the form (6.30). This implies that

$$a^{(1)}_{\mu_1}\dots a^{(N-1)}_{\mu_{N-1}}\,\frac{T^{\mu_1\dots\mu_N}(\xi)}{\xi^{\mu_N}}\quad\text{, no summation over } \mu_N ,$$

is an invariant distribution and does not depend on μ_N . (Distributions in $\mathscr{S}'_s(\xi)$ can be divided by powers of ξ without problem, because the test functions from $\mathscr{S}_s$ vanish strongly in $\xi=0$.) But then

$$\frac{T^{\mu_1\dots\mu_N}(\xi)}{\xi^{\mu_N}} = T^{\mu_1\dots\mu_{N-1}}(\xi) \tag{6.32}$$

is a contravariant tensor of rank $N-1$ [30] and is of the form

$$T^{\mu_1\dots\mu_{N-1}}(\xi) = \xi^{\mu_1}..\xi^{\mu_{N-1}}T(\xi^2) \tag{6.33}$$

by the induction hypothesis. (6.32) and (6.33) give immediately the desired representation (6.15).

From $T(\xi) = T(\xi^2)$ we learn that

$$T(\xi) = T(-\xi) : \tag{6.34}$$

any distribution $T(\xi) \in \mathscr{S}'_S(\xi)$ which is invariant under $L_+^\uparrow$ is automatically also invariant under the PT-operation $\xi \to -\xi$ [28,29] . With this fact in mind we obtain from (6.15):

$$T^{\mu_1 \cdots \mu_N}(-\xi) = (-1)^N \, T^{\mu_1 \cdots \mu_N}(\xi) \quad . \tag{6.35}$$

Application of this result to (6.13) yields

$$S(-x,-y,-X) = S(x,y,X) \quad . \tag{6.36}$$

The remainder of the chapter will be devoted to the proof of the CTP relation

$$S(-x,-y,-X) = - \, S(x,y,X) \quad . \tag{6.37}$$

Comparison of (6.36) and (6.37) then gives the desired result

$$S(x,y,X) = 0 \quad . \tag{6.38}$$

The proof of (6.37) cannot be given in the accustomed fashion, by discussing the properties of the c-number distributions r_σ and I_σ only. It necessitates a sizable detour through the Hilbert space $\mathfrak{H}$. As a reward of this labour we obtain, in addition to (6.37), a general proof of the CTP invariance of our formalism.

We shall need the g.r.f. g_μ , and thus we must briefly discuss how their expansion coefficients $g_{\mu,\sigma}$ can be calculated from r_σ . The perturbative form of the completeness equation (2.68) is

$$g_{\mu,\sigma}(X) - g_{\nu,\sigma}(X) = -i \sum_{\ell=1}^{\infty} \frac{1}{\ell!} \sum_{\tau=1}^{\sigma-1} \int \prod_1^{\ell}\{du_i\, dv_i\}\ K^{\ell}(U-V)$$

$$\times\ g^+_{\alpha,\tau}(X_L;U)\ g^+_{\beta,\sigma-\tau}(X_R;V) \quad . \qquad (6.39)$$

We assume all $g_{\mu,\tau}$ known in all orders $\tau < \sigma$. Then we can calculate $g_{\mu,\sigma}$ from (6.39) if we know $g_{\nu,\sigma}$. Starting from the known r_σ we can in this way calculate successively all the $g_{\mu,\alpha}$ (see the considerations leading to (2.63)) as r_σ plus a number of right-hand sides of equations (6.39). The existence of the integrals in (6.39) will be proved in Chapter VII. The $g_{\mu,\sigma}$ turn out to be real. With the help of the expansion (2.69) we can then define the g.r.p.

$$G_{\mu,\sigma}(X) = g_{\mu,\sigma}(X) + \sum_{\ell=1}^{\infty} \frac{1}{\ell!} \int du_1..du_\ell\ g^+_{\mu,\sigma}(X;U)\ \ :A^{in}(u_1)...A^{in}(u_\ell): \quad . \qquad (6.40)$$

Note that only a finite number of $g^+_{\mu,\sigma}$ are different from zero in any finite order σ , so that the ℓ-sum runs only over a finite number of terms.

In Chapter II (see eq.(2.40)) we introduced the space $\mathcal{L}^{in}_\varepsilon$ of in-states with a finite number of particles with Hölder continuous wave functions from H_ε . In the next chapter $g^+_{\mu,\sigma}$ will be shown to possess property iii) of Theorem 2.1 . From this follows that $G_{\mu,\sigma}$ is defined on $\mathcal{L}^{in}_\varepsilon$ and its integral over a test function $\phi(X)$ maps $\mathcal{L}^{in}_\varepsilon$ into $\mathcal{L}^{in}_{\varepsilon'}$, $\varepsilon' < \varepsilon$.

From (2.44) and (2.50) we obtain as perturbation coefficients of the field A

$$A_\sigma(x) = \sum_{\ell=1}^{\infty} \int \frac{1}{\ell!} \, du_1 \ldots du_\ell \, dy \, \Delta_R(x-y) \, r_\sigma(y,U) \, :A^{in}(u_1)..A^{in}(u_\ell): \; . \tag{6.41}$$

We prove

Theorem 6.2

The following statements are true.

1) $$\sum_{\tau=0}^{\sigma} [A_\tau(x), A_{\sigma-\tau}(y)] = 0 \quad \text{for} \quad (x-y)^2 < 0 \; , \tag{6.42}$$

$$\sum_{\tau=0}^{\sigma} [G_{\mu,\tau}(X), G_{\nu,\sigma-\tau}(y)] = 0 \text{ if } (x_i - y_j)^2 < 0 \text{ for all } x_i \in X, \; y_j \in Y \; . \tag{6.43}$$

2) There exist antilinear operators Θ_τ which are defined on $\mathcal{L}_\varepsilon^{in}$, map $\mathcal{L}_\varepsilon^{in}$ into itself, and satisfy

$$\sum_{\tau=0}^{\sigma} \Theta_\tau \, \Theta_{\sigma-\tau} = \begin{cases} 1 & \text{for } \sigma = 0 \\ 0 & \text{for } \sigma > 0 \; , \end{cases} \tag{6.44}$$

$$\sum_{\tau=0}^{\sigma} (\Theta_\tau \Phi, \Theta_{\sigma-\tau} \Psi) = \begin{cases} (\Phi,\Psi)^* & \text{for } \sigma = 0 \; , \\ 0 & \text{for } \sigma > 0 \; , \end{cases} \quad \Phi, \Psi \in \mathcal{L}_\varepsilon^{in} \tag{6.45}$$

$$\sum_{\tau'+\tau''+\tau'''=\sigma} \Theta_{\tau'} G_{\mu,\tau''}(x_1,..,x_n) \Theta_{\tau'''} = \overline{G}_{\mu,\sigma}(-x_1,..,-x_n) \; , \tag{6.46}$$

$$\Theta_\sigma \Omega = \begin{cases} \Omega & \text{for } \sigma = 0 \\ 0 & \text{for } \sigma > 0 \; . \end{cases} \tag{6.47}$$

For the definition of $\overline{G}_\mu$ we refer to Chapter II.

Point 1) of the theorem states the locality of $A(x)$ and of the amputated g.r.p. Point 2) claims the CTP-invariance of the theory.

The Θ_τ are the expansion coefficients of the CTP operator Θ introduced in Chapter II. (6.44) is the perturbative form of (2.72), (6.46) of (2.76), (6.47) of (2.73), and (6.45) states the anti-unitarity of Θ.

We prove the theorem by induction with respect to σ. The theorems holds obviously in order zero, with

$$A_o(x) = A^{in}(x) \tag{6.48}$$

and Θ_o the CTP operator of the free field $A^{in}(x)$, so that

$$\Theta_o A_o(x)\Theta_o = A_o(-x) \quad , \tag{6.49}$$

$$\Theta_o^{\;2} = 1 \quad , \tag{6.50}$$

$$\Theta_o \Omega = \Omega \quad . \tag{6.51}$$

We assume that the theorem is true in all orders $\tau < \sigma$. From (6.46) and (6.47) we obtain

$$\bar{g}_{\mu,\tau}(X) = g_{\mu,\tau}(-X) \quad . \tag{6.52}$$

Another consequence of the assumed validity of the theorem in lower orders is that

$$A_\tau^{out}(x) = \sum_{\tau'=0}^{\tau} \Theta_{\tau'} A^{in}(-x)\Theta_{\tau-\tau'} \tag{6.53}$$

are the expansion coefficients of a free field, i.e.

$$\sum_{\tau'=0}^{\tau} [A_{\tau'}^{out}(x), A_{\tau-\tau'}^{out}(y)] = \begin{cases} i\Delta(x-y) & \text{for } \tau=0 \\ 0 & \text{for } 0<\tau<\sigma \quad . \end{cases} \tag{6.54}$$

The $G_{\mu,\tau}$, $\tau < \sigma$, can be expanded with respect to A^{out}. We start from the expansion (6.40) for $\bar{G}_{\mu,\tau}$:

$$\bar{G}_{\mu,\tau}(-X) = \sum_{\ell=0}^{\infty} \frac{1}{\ell!} \int dU\, \bar{g}^{+}_{\mu,\tau}(-X;U)\ :A^{in}(u_1)\ldots A^{in}(u_\ell): \quad ,$$

hence, using the reality of $\bar{g}^{+}_{\mu,\tau}$,

$$\Sigma_{\tau'+\tau''+\tau'''=\tau}\ \Theta_{\tau'}\bar{G}_{\mu,\tau''}(-X)\Theta_{\tau'''} = \Sigma_{\ell=0}^{\infty} \frac{1}{\ell!}\ \Sigma_{\tau'+\tau''+\tau'''=\tau}$$

$$\times \int dU\, g^{+}_{\mu,\tau'}(-X;U)\ \Theta_{\tau''}:\ldots A^{in}(u_i)\ldots:\ \Theta_{\tau'''} \quad .$$

This becomes by applying successively (6.46), (6.44), (6.53), and (6.52)

$$\begin{aligned} G_{\mu,\tau}(X) &= \sum_{\ell} \frac{1}{\ell!} \sum_{\tau'+\tau''=\tau} \int dU\, g^{+}_{\mu,\tau'}(-X;U)\ (:\prod_{1}^{\ell} A^{out}(-u_i):)_{\tau''} \\ &= \sum_{\ell} \frac{1}{\ell!} \sum_{\tau'+\tau''=\tau} \int dU\, g^{-}_{\mu,\tau'}(X;U)\ (:\prod_{1}^{\ell} A^{out}(u_i):)_{\tau''} \quad . \end{aligned} \tag{6.55}$$

Consider

$$H_{\mu\nu,\sigma}(X,Y) = \sum_{\tau=1}^{\sigma-1} [G_{\mu,\tau}(X), G_{\nu,\sigma-\tau}(Y)] \quad , \tag{6.56}$$

with $X = \{x_1,\ldots,x_n\}$, $Y = \{y_1,\ldots,y_m\}$, at the points where X is totally space-like with respect to Y , i.e. all x_i are space-like to all y_j . The $\tau = 0, \sigma$ terms do not occur because $G_{\nu,o}(Y)$ vanishes for $m \neq 2$ and is a c-number for $m=2$, due to (3.5). Let $G_{\mu'}(X)$, $G_{\mu''}(X)$ belong to two adjacent cells $C_{\mu',\mu''}$, so that

$$G_{\mu'}(X) - G_{\mu''}(X) = -i\,[G_\alpha(X_L), G_\beta(X_R)]$$

with $X_{L,R}$ two complementary subsets of X . We obtain with the help of the Jacobi identities

$$\begin{aligned} H_{\mu'\nu,\sigma} - H_{\mu''\nu,\sigma} = i \sum_{\tau=1}^{\sigma-1} \sum_{\tau'=1}^{\tau-1} \{ &[[G_{\beta,\tau-\tau'}(X_R), G_{\nu,\sigma-\tau}(Y)], G_{\alpha,\tau'}(X_L)] \\ &+ [[G_{\nu,\sigma-\tau}(Y), G_{\alpha,\tau-\tau'}(X_L)], G_{\beta,\tau'}(X_R)] \} \quad . \end{aligned}$$

But $\sigma-\tau'<\sigma$, so that the two inner commutators vanish by the induction hypothesis:

$$H_{\mu'\nu,\sigma}(X,Y) = H_{\mu''\nu,\sigma}(X,Y) \tag{6.57}$$

for X totally space-like to Y . Since any two $C_{\mu'}$, $C_{\mu''}$ can be connected by a chain of adjacent cells we find that $H_{\mu\nu,\sigma}(X,Y)$ does not, in these points, depend on the particular g.r.p. G_μ chosen. The same is true for the second factor G_ν , so that

$$H_{\mu\nu,\sigma}(X,Y) = \sum_{\tau=1}^{\sigma-1} [R_\tau(X),R_{\sigma-\tau}(Y)] = \sum_{\tau=1}^{\sigma-1} [\bar{R}_\tau(X),\bar{R}_{\sigma-\tau}(Y)] \quad . \tag{6.58}$$

But $R_\tau(x_1,..x_n)$, $\tau<\sigma$, has the support $(x_1-x_2)\in\bar{V}_+$, .. , $(x_1-x_n)\in\bar{V}_+$, $\bar{R}_\tau$ the support $(x_1-x_2)\in\bar{V}_-$, .. , $(x_1-x_n)\in\bar{V}_-$, hence $H_{\mu\nu\sigma}$ has its support contained in $x_1=...=x_n$ and analogously in $y_1=...=y_m$. This is a generalisation of our earlier result on the support of $I_{L\sigma}$.

We introduce

$$h_{\mu\nu,\sigma}(X,Y) = \langle 0|H_{\mu\nu,\sigma}(X,Y)|0\rangle \quad . \tag{6.59}$$

For the moment we drop the restriction to relatively space-like X , Y and calculate $h_{\mu\nu,\sigma}(X,Y)$ by inserting the expansions (6.40) for $G_{\mu,\tau}$, $G_{\nu,\sigma-\tau}$. The products :...: :...: can be reduced to sums over Wick products by Wick's theorem. We obtain the usual completeness expressions such as the right-hand side of (6.39). (After all, the $g_{\mu,\sigma}$ have been constructed such that this is true.) They read

$$h_{\mu\nu,\sigma}(X,Y) = -i\sum_{\ell=1}^{\infty}\frac{1}{\ell!}\sum_{\tau=1}^{\sigma-1}\int dU\,dV\,K^\ell(U-V)\,g^+_{\mu,\tau}(X;U)\,g^+_{\nu,\sigma-\tau}(Y;V) \quad . \tag{6.60}$$

If we use the out-expansion (6.55) we get in the same way

$$h_{\mu\nu,\sigma}(X,Y) = -i \sum_{\ell} \frac{1}{\ell!} \sum_{\tau=1}^{\sigma-1} \int dU\, dV\, K^{\ell}(U-V)\, g^{-}_{\mu,\tau}(X;U)\, g^{-}_{\nu,\sigma-\tau}(Y;V) \quad .$$

Using (6.52) and the antisymmetry of K^{ℓ} : $K^{\ell}(U-V) = -K^{\ell}(-U+V)$, we obtain

$$h_{\mu\nu,\sigma}(X,Y) = i \sum_{\ell} \frac{1}{\ell!} \sum_{\tau=1}^{\sigma-1} \int dU\, dV\, K^{\ell}(U-V)\, \bar{g}^{+}_{\mu,\tau}(-X;U)\, \bar{g}^{+}_{\nu,\sigma-\tau}(-Y;V) \quad ,$$

and this is,according to (6.60),

$$h_{\mu\nu,\sigma}(X,Y) = - \sum_{\tau=1}^{\sigma-1} \langle 0|[\bar{G}_{\mu,\tau}(-X),\bar{G}_{\nu,\sigma-\tau}(-Y)]|0\rangle. \qquad (6.61)$$

If we now restrict ourselves again to relatively space-like X and Y we obtain by using (6.58)

$$\sum_{\tau} \langle 0|[R_{\tau}(X),R_{\sigma-\tau}(Y)]|0\rangle = - \sum_{\tau} \langle 0|[R_{\tau}(-X),R_{\sigma-\tau}(-Y)]|0\rangle \quad , \qquad (6.62)$$

and this is, if expressed in the form (6.60), the desired relation (6.37).

With this we have provided the missing link in the construction of r_{σ} : the proof of the validity of the support condition a) of Theorem 4.2 . Hence r_{σ} can be calculated as proposed and has all the desired properties, in particular the correct support. $G_{\mu,\sigma}$ and A_{σ} can then be calculated from (6.40) and (6.41). $R_{\sigma}(x_1,..,x_n)$ as defined by (6.40) has the support $(x_1-x_i)\in\bar{V}_{+}$, $i=2,..,n$, and satisfies the relation

$$R_{\sigma}(x,y,X) - R_{\sigma}(y,x,X) = -i \sum_{L,R} \sum_{\tau=1}^{\sigma-1} [R_{\tau}(x,X_L), R_{\sigma-\tau}(y,X_R)] \quad .(6.63)$$

This can be proved in the same way as (2.46) was proved in Chapter II. As usual, X_L and X_R are complementary subsets of $X = \{x_1,...,x_n\}$.

Consider (6.63) at the points in which for a particular partition $X = X_L \cup X_R$ the two sets $\{x,X_L\}$ and $\{y,X_R\}$ are totally space-like to each other. There all the terms on the right-hand side of (6.63) vanish, except possibly the term corresponding to the particular partition under consideration (see (6.58)). But the left-hand side vanishes because of the support of R_σ , hence the critical L-term vanishes too. This proves (6.43) for the special case $G_\mu = R$, $G_\nu = R$,and thus in general because of (6.58).

It has been remarked in connection with the equations (2.50) that de-amputation of r , and by the same token also of R ,leaves the support invariant. Hence (6.42) follows from the support of $R_\sigma(x,y)$ and the identity

$$[A(x),A(y)]_\sigma = R_\sigma^{deamp}(x,y) - R_\sigma^{deamp}(y,x) \quad . \tag{6.64}$$

We turn now the CTP part of Theorem 6.2.

A representation of the form (2.64) can be derived, in particular, for $\bar{R}(x,y,x_1,..,x_n)$. After some simple rearrangements it becomes

$$\bar{R}(x,y,X) = R(y,x,X) + \Sigma\, i^{\ell-1} c_{\alpha_1 \cdots \alpha_\ell} [\ldots[R(y,x,X_{\alpha_1}),R(X_{\alpha_2})],..,R(X_{\alpha_\ell})]. \tag{6.65}$$

As do all these algebraic relations, (6.65) holds in order σ of perturbation theory. From the known supports of R_τ we deduce that $\bar{R}_\sigma$ vanishes outside of $(x-y)\in\bar{V}_-$. Because of the symmetry of $\bar{R}_\sigma(x,x_1,..,x_n)$ in $x_1,..,x_n$ (which is again one of those valid algebraic relations) the support of $\bar{R}_\sigma(x,X)$, and therefore also of $\bar{r}_\sigma(x,X)$, lies in $(x-x_i)\in\bar{V}_-$, all i .

A special case of relation (2.62) gives

$$\begin{aligned}\bar{R}_\sigma(X) &= R_\sigma(X) + i \sum_{\alpha,\beta} c_{\alpha\beta} \sum_{\tau=1}^{\sigma-1} [G_{\alpha,\tau}(X_\alpha),G_{\beta,\sigma-\tau}(X_\beta)] \\ &= R_\sigma(X) + i \sum_{\alpha\ \beta} c_{\alpha\beta} \sum_{\tau=1}^{\sigma-1} [G_{\alpha,\tau}(X_\alpha),G_{\beta,\sigma-\tau}(X_\beta)]\end{aligned} \tag{6.66}$$

with $c_{\alpha\beta}$ real. Taking the vacuum expectation value and using the known properties of the operators Θ_τ , $\tau < \sigma$, (eq. (6.44) to (6.47)) we obtain

$$\bar{r}_\sigma(-X) = r_\sigma(-X) + i \sum_{\alpha,\beta} c_{\alpha\beta} \sum_\tau \langle 0|[G_{\alpha\tau}(-X_\alpha), G_{\beta\,\sigma-\tau}(-X_\beta)]|0\rangle$$

$$= r_\sigma(-X) - i \sum_{\alpha,\beta} c_{\alpha\beta} \sum_\tau \langle 0|[\bar{G}_{\alpha\tau}(X_\alpha), \bar{G}_{\beta,\sigma-\tau}(X_\beta)]|0\rangle$$

$$= r_\sigma(-X) + r_\sigma(X) - \bar{r}_\sigma(X) \quad ,$$

whence

$$M(X) = r_\sigma(X) - \bar{r}_\sigma(-X) = - r_\sigma(-X) + \bar{r}_\sigma(X) = - M(-X) \quad . \quad (6.67)$$

$M(X)$ has its support in $(x_1-x_i)\epsilon\bar{V}_+$, $i=2,..,n$, so that (6.67) implies the support $x_1=x_2=...=x_n$ for M . Furthermore M is Poincaré invariant, hence

$$M(X) = D \prod_{i=2}^{n} \delta^4(x_1-x_i) \quad ,$$

with D an invariant differential operator. But this implies $M(X) = M(-X)$ in contradiction with (6.67) unless $M(X) = 0$:

$$r_\sigma(X) = \bar{r}_\sigma(-X) \quad . \quad (6.68)$$

From (6.68), the perturbative form of (2.63):

$$G_{\mu\sigma}(X) = R_\sigma(X) - i \sum_{\alpha,\beta} c^\mu_{\alpha\beta} [G_\alpha(X_\alpha), G_\beta(X_\beta)]_\sigma \quad , \quad (6.69)$$

and its CTP transformed

$$\bar{G}_{\mu\sigma}(X) = \bar{R}_\sigma(X) + i \sum_{\alpha,\beta} c^\mu_{\alpha\beta} [\bar{G}_\alpha(X_\alpha), \bar{G}_\beta(X_\beta)]_\sigma \quad (6.70)$$

we obtain with the help of (6.61) the more general relation

$$g_{\mu,\sigma}(X) = \bar{g}_{\mu,\sigma}(-X) \quad ,$$

i.e. equation (6.52) in order σ .

Finally we prove the existence of Θ_σ. On Ω we define Θ_σ according to (6.47):

$$\Theta_\sigma \Omega = 0 \quad . \tag{6.71}$$

More generally we define

$$\Theta_\sigma A_0(x_1) \ldots A_0(x_n) \Omega$$

$$= \sum_{\Sigma\sigma_i=\sigma} A_{\tau_1}(-x_1) \ldots A_{\tau_n}(-x_n)\Omega - \sum_{\tau=1} \sum_{\Sigma\tau_i=\tau} \Theta_{\sigma-\tau} A_{\tau_1}(x_1) \ldots\ldots A_{\tau_n}(x_n)\Omega \tag{6.72}$$

This definition is made such that

$$[\Theta A(x_1)\ldots A(x_n)\Theta]_\sigma \Omega = [A(-x_1)\ldots A(-x_n)]_\sigma \Omega \quad . \tag{6.73}$$

We introduce the notation

$$A_\tau(X) = \sum_{\Sigma\tau_i=\tau} A_{\tau_1}(x_1)\ldots A_{\tau_n}(x_n) \tag{6.74}$$

and calculate with the help of (6.50)and the induction hypothesis $\sum_{\tau'=0}^{\tau} \Theta_{\tau'}\Theta_{\tau-\tau'} = 0$ for $0<\tau<\sigma$:

$$\sum_{\tau=0}^{\sigma} \Theta_{\sigma-\tau}\Theta_\tau A_0(X)\Omega = \sum_{\tau=1}^{\sigma} \Theta_{\sigma-\tau}A_\tau(-X)\Omega - \sum_{\tau=1}^{\sigma} \sum_{\tau'=1}^{\tau} \Theta_{\sigma-\tau}\Theta_{\tau-\tau'} A_{\tau'}(X)\Omega$$

$$= A_\sigma(X)\Omega - \sum_{\tau'>0} \sum_{\tau_1+\tau_2=\sigma-\tau'} \Theta_{\tau_1}\Theta_{\tau_2} A_{\tau'}(X)\Omega$$

$$= A_\sigma(X)\Omega - A_\sigma(X)\Omega$$

$$= 0 \quad ,$$

which proves (6.44) in order σ .

Let $B_\tau(Y)$ be of the form (6.74):

$$B_\tau(Y) = \sum_{\Sigma\tau_i = \tau} A_{\tau_1}(y_1)\dots A_{\tau_m}(y_m) \quad . \tag{6.75}$$

We know already that locality is satisfied in order σ, hence the Wightman functions

$$W_\sigma(x_1,\dots,x_n,y_1,\dots,y_m) = \sum_{\tau=0}^{\sigma} \langle 0|A_\tau(X)\; B_{\sigma-\tau}(Y)|0\rangle \tag{6.76}$$

are local. They are also invariant and have the correct spectral properties, so that Jost's CTP theorem [13,14] holds:

$$W_\sigma(X,Y) = W_\sigma(-\overleftarrow{Y},-\overleftarrow{X}) \quad . \tag{6.77}$$

$\overleftarrow{X}$, $\overleftarrow{Y}$, contain the variables of X, Y, in reversed order. From (6.77) and (6.73) we deduce

$$\begin{aligned}(A(X)\Omega\;,B(Y)\Omega)_\sigma &= (B(-\overleftarrow{Y})\Omega\;,A(-\overleftarrow{X})\Omega)_\sigma \\ &=(\Theta B(\overleftarrow{Y})\Omega\;,\Theta A(\overleftarrow{X})\Omega)_\sigma \\ &= \sum_{\Sigma\tau_i=\sigma} (\Theta_{\tau_1}\; B_{\tau_2}(\overleftarrow{Y})\Omega\;,\Theta_{\tau_3}\; A_{\tau_4}(\overleftarrow{X})\Omega) \quad .\end{aligned}$$

For $\tau_2+\tau_4 \neq 0$ we can sum over τ_1, τ_3 with the help of assumption (6.45) and obtain

$$\begin{aligned}\sum_\tau (\Theta_\tau B_o(\overleftarrow{Y})\Omega\;,\Theta_{\sigma-\tau}A_o(\overleftarrow{X})\Omega) + \sum_\tau (A_\tau(X)\Omega\;,B_{\sigma-\tau}(Y)\Omega) \\ = \sum_\tau (A_\tau(X)\Omega\;,B_{\sigma-\tau}(Y)\Omega) \quad ,\end{aligned}$$

or

$$\sum_\tau (\Theta_\tau B_o(\overleftarrow{Y})\Omega\;,\Theta_{\sigma-\tau}A_o(\overleftarrow{X})\Omega) = 0 \quad ,$$

which proves (6.45) in order σ .

The results obtained so far suffice to prove (6.54) and (6.55) in order σ. Hence, using (6.52) (this has already been shown to be valid in order σ):

$$[\Theta G_\mu(X)\Theta]_\sigma = \sum_\ell \frac{1}{\ell!} \sum_{\tau'+\tau''+\tau'''=\sigma} \int dU\; g^+_{\mu,\tau'}(X;U)\; \Theta_{\tau''}\colon \prod_{i=1}^{\ell} A^{in}(u_i)\colon \Theta_{\tau'''}$$

$$= \sum_\ell \frac{1}{\ell!} \sum_{\tau'+\tau''=\sigma} \int dU\; g^+_{\mu,\tau'}(X;-U)\; \Big[\colon \prod_1^{\ell} A^{out}(u_i)\colon\Big]_{\tau''}$$

$$= \sum_\ell \frac{1}{\ell!} \sum_{\tau'+\tau''=\sigma} \int dU\; \bar{g}^-_{\mu,\tau'}(-X;U)\; \Big[\colon \prod_1^{\ell} A^{out}(u_i)\colon\Big]_{\tau''}$$

$$= \bar{G}_{\mu,\sigma}(-X) \quad ,$$

which proves (6.46) in order σ . This completes the proof of Theorem 6.2.

VII. EXISTENCE

We must still solve the long postponed problem of the existence of I_σ . In order to do this we prove that the solutions $\tilde{r}_\sigma$ as constructed in the earlier chapters satisfy condition iii) of Theorem 2.1. Or rather, we prove a stronger property.

We come back to the definition of H_ε given in Chapter II and generalise it slightly. Let $f(u_1,\ldots,u_n,v_1,\ldots,v_m)$ be Hölder continuous of index ε in all variables. We say that f is strongly decreasing in U and slowly increasing in V if

$$S_N(f|V) = \sup_{\substack{U \\ |A| \leq A_0}} \{ (1+|U|^2)^N [|f(u_1,\ldots,v_m)| + \frac{|f(u_1+a_1,\ldots,u_n+a_n,v_1+a_{n+1},\ldots,v_m+a_{n+m}) - f(u_1,\ldots v_m)|}{|A|^\varepsilon}]\} \tag{7.1}$$

exists for all non-negative integers N and is polynomially bounded in V. A_0 is an arbitrary but fixed positive constant. Remember $|A|^2=\Sigma|a_i|^2$. The space of the functions satisfying this condition will be called $H_\varepsilon(U;V)$. For $V=\emptyset$ this is simply the original $H_\varepsilon(U)$.

The main result of this chapter is

Theorem 7.1

The $\tilde{g}_{\mu\sigma}$ as constructed in Chapters III-VI exist as tempered distributions. Moreover, if $f(k_1,\ldots,k_\alpha,p_1,\ldots,p_\beta,q_1,\ldots,q_\gamma,u_1,\ldots,u_\delta,v_1,\ldots,v_\zeta) \in H_\varepsilon(K,\underset{\sim}{P},U;P^-,Q,V)$, $\varepsilon \leq \frac{1}{2}$, then

$$\mathcal{F}(P^-,Q,U,V) = \prod_{i=1}^{\gamma} \chi(-q_i) \int dK\, d\underset{\sim}{P} \prod_{i=1}^{\beta} \chi(p_i)\; \tilde{g}_{\mu\sigma}(K,P,Q)\, f(K,P,Q,U,V) \tag{7.2}$$

exists and is in $H_{\varepsilon'}(P^-,Q,U;V)$ for any $\varepsilon'<\varepsilon$. The case $\alpha=0$, $\beta+\gamma=2$, is excluded.

For the definition of χ see Chapter II, Eq. (2.41) and (2.42). The p_i-dependence of all the factors in the integrand of (7.2) must be written as dependence on $\underset{\sim}{p}_i$ and $p_i^- = p_{io} - \omega(\underset{\sim}{p}_i)$ rather than $\underset{\sim}{p}_i$ and p_{io}. In other words: the $\underset{\sim}{p}_i$-integration must be carried out with p_i^-, not p_{io}, being kept constant. P^- is the set $\{p_1^-,\ldots,p_\beta^-\}$.

The order in which the various variables appear in $\tilde{g}_{\mu,\sigma}$ is irrelevant, the k_i need not necessarily stand in front, etc. The theorem holds, of course, also if the variables P and Q exchange their roles, so that we integrate over the variables lying near the negative mass shell.

We prove the theorem by induction with respect to σ . As the start of the induction we prove first that the theorem holds in <u>first order</u>.

All the $g_{\mu,1}(x_1,..,x_n)$ with the same number n of arguments are equal and are given by (4.6). In a theory of type (μ,ν) only the μ-point functions are different from zero and are, in p-space,

$$\tilde{g}_{\mu 1}(p_1,..,p_\mu) = \delta^4(p_1+..+p_\mu)\ \mathcal{P}(p_1,..,p_\mu) \qquad (7.3)$$

with $\mathcal{P}$ a totally symmetric, invariant form of degree ν . We must prove that this distribution has the property claimed in Theorem 7.1. We distinguish two cases.

<u>1st case: $\alpha > 0$</u> . We can perform the k_1-integration with the help of the δ^4-factor in (7.2). The result is

$$\mathcal{F}(P^-,Q,U,V) = \int d\hat{K}\, d\underset{\approx}{P}\ f_o(\hat{K},P,Q,U,V) \qquad (7.4)$$

with $\hat{K} = \{k_2,..,k_\alpha\}$ and

$$f_o(\hat{K},P,Q,U,V) = \prod_Q \chi(-q_i) \prod_P \chi(p_j) \mathcal{P}(K,P,Q)\ f(K,P,Q,U,V)\Big|_{k_1=-\sum_2^\alpha k_i-\sum_1^\beta p_j-\sum_1^\gamma q_h} . \qquad (7.5)$$

From the support of χ we infer that $f_o \in H_\varepsilon(\hat{K},P^-,\underset{\approx}{P},Q,U;V)$. Hence $\mathcal{F}$ exists, is continuous everywhere and decreases for fixed V strongly in the remaining variables. We define

$$h^2 = \sum_j (\Delta p_j)^2 + \sum_j |\Delta q_j|^2 + \sum_j |\Delta u_j|^2 + \sum_j |\Delta v_j|^2$$

and estimate

$$|\bar{f}(P^- + \Delta P^-, Q+\Delta Q, U+\Delta U, V+\Delta V) - \bar{f}(P^-, Q, U, V)|$$

$$\leqslant \int d\hat{K}\, d\underset{\sim}{P}\, |f_o(\hat{K}, \underset{\sim}{P}, P^- + \Delta P^-, Q+\Delta Q, U+\Delta U, V+\Delta V) - f_o(\hat{K}, \ldots, V)|$$

$$\leqslant h^\varepsilon\, S_{N+M}(f_o|V)\, (1+|P^-|^2+|Q|^2+|U|^2)^{-N} \int d\hat{K}\, d\underset{\sim}{P}\, (1+|\hat{K}|^2+|\underset{\sim}{P}|^2)^{-M} \quad .$$

The last integral exists for M sufficiently large. This proves that $\bar{f}\in H_\varepsilon(P^-, Q, U; V)$, q.e.d.

<u>2nd case: $\alpha=0$</u> . In this case the theorem makes no statement on the 2-point function. The 3-point function vanishes because of the support (2.41) of χ . So do all the higher functions unless $\beta \geqslant 2$ and $\gamma \geqslant 2$. Eq. (7.2) can then be written

$$\bar{f}(P^-, Q, U, V) = \int d\underset{\sim}{p}_1\, d\underset{\sim}{p}_2\, d\hat{\underset{\sim}{P}}\, f_o(\underset{\sim}{P}, P^-, Q, U, V)\, \delta^3\Big(\sum_1^\beta \underset{\sim}{p}_i + \sum_1^\gamma \underset{\sim}{q}_i\Big)$$

$$\times\ \delta\Big(\sum_1^\beta p_i^- + \sum_1^\beta \omega(\underset{\sim}{p}_i) + \sum_1^\gamma q_{io}\Big)$$

$$= \int d\underset{\sim}{p}_2\, d\hat{\underset{\sim}{P}}\, f_o(\underset{\sim}{P}, P^-, Q, U, V)\, \delta(\Omega + \omega(\underset{\sim}{p}_1) + \omega(\underset{\sim}{p}_2)) \quad , \tag{7.6}$$

where $\hat{P} = \{p_3, \ldots, p_\beta\}$,

$$f_o(\underset{\sim}{P}, \ldots, V) = \prod_1^\gamma \chi(-q_i) \prod_1^\beta \chi(p_i)\, \Phi(\underset{\sim}{P}, P^-, Q)\, f(\underset{\sim}{P}, P^-, Q, U, V) \quad , \tag{7.7}$$

and

$$\underset{\sim}{p}_1 = -\underset{\sim}{p}_2 - \underset{\sim}{p} \quad , \quad \underset{\sim}{p} = \sum_3^\beta \underset{\sim}{p}_i + \sum_1^\gamma \underset{\sim}{q}_i \quad , \tag{7.8}$$

$$\Omega = \sum_1^\beta p_i^- + \sum_3^\beta \omega(\underset{\sim}{p}_i) + \sum_1^\gamma q_{io} \quad . \tag{7.9}$$

The $\underset{\sim}{p}_2$-integration can be performed easily in the special case $\underset{\sim}{p}=0$, which implies $\omega(\underset{\sim}{p}_1) = \omega(\underset{\sim}{p}_2)$. Integration of f_o over angular variables in $\underset{\sim}{p}_2$ yields a H_ε-function $F(\omega_2,\hat{\underset{\sim}{P}},P^-,Q,U,V)$. The $|\underset{\sim}{p}_2|$-integration can be transformed into $\int_{\omega_2 \geq m} \omega_2\sqrt{\omega_2^2-m^2}\,d\omega_2$ and can be carried out with the remaining δ-function. The result is

$$-\frac{\Omega}{4}\,\theta(\Omega^2-4m^2)\,\sqrt{\Omega^2-4m^2}\quad F(-\frac{\Omega}{2},\hat{\underset{\sim}{P}},P^-,Q,U,V) \quad . \tag{7.10}$$

Under our assumption $\varepsilon \leqslant \frac{1}{2}$ this is in $H_\varepsilon(\hat{\underset{\sim}{P}},P^-,Q,U;V)$, because the product of two Hölder continuous functions of index ε is again Hölder continuous of the same index.

This behaviour also occurs for arbitrary $\underset{\sim}{p}$. In this case $\omega(\underset{\sim}{p}_1) + \omega(\underset{\sim}{p}_2)$ is stationary at $\underset{\sim}{p}_2 = -\underset{\sim}{p}/2$. For Ω not corresponding to this critical value the $\underset{\sim}{p}_2$-integration does not introduce any singularities besides the ones present in f . For Ω in the vicinity of the critical value $\bar{\Omega}(\underset{\sim}{p}) = -2\omega(\underset{\sim}{p}/2)$ we can develop

$$\omega(\underset{\sim}{p}_1) + \omega(\underset{\sim}{p}_2) = -\bar{\Omega}(\underset{\sim}{p}) + A_2(\underset{\sim}{p},\underset{\sim}{p}_2+\frac{\underset{\sim}{p}}{2}) + \text{higher terms in } (\underset{\sim}{p}_2+\frac{\underset{\sim}{p}}{2}) \,, \tag{7.11}$$

with

$$A_2(\underset{\sim}{p},\underset{\sim}{q}) = 2(\underset{\sim}{p}^2+4m^2)^{-\frac{3}{2}}\,\{(\underset{\sim}{p}^2+4m^2)\underset{\sim}{q}^2 - (\underset{\sim}{p},\underset{\sim}{q})^2\} \quad . \tag{7.12}$$

A_2 is for fixed $\underset{\sim}{p}$ a positive definite quadratic form in $\underset{\sim}{q} = \underset{\sim}{p}_2 + \underset{\sim}{p}/2$. For performing the $\underset{\sim}{p}_2$-integration we chose a coordinate system such that $\underset{\sim}{p}$ lies in the 3-direction. After the substitution

$$\begin{aligned} q_1 &= t_1\,(\underset{\sim}{p}^2+4m^2)^{1/4} \\ q_2 &= t_2\,(\underset{\sim}{p}^2+4m^2)^{1/4} \\ q_3 &= t_3\,\frac{(\underset{\sim}{p}^2+4m^2)^{3/4}}{2m} \end{aligned} \tag{7.13}$$

we obtain for the $\underset{\sim}{p}_2$-integral in (7.6)

$$\frac{(\underset{\sim}{p}^2+4m^2)^{5/4}}{2m} \int d^3t \;\; \delta(\Omega - \overline{\Omega} + 2\, t^2 + \text{higher terms in } t\,)$$

$$\times \;\; f'(t,\hat{\underset{\sim}{P}},P^-,Q,U,V) \qquad (7.14)$$

with $f' \in H_\varepsilon(\ldots;V)$ .The higher terms in the argument of δ do not influence the nature of the singularity in $\Omega=\overline{\Omega}$ and can be dropped*. The integration is again performed by integrating first over angular variables in $\underset{\sim}{t}$ and then over $|\underset{\sim}{t}|$. We obtain

$$\theta(\overline{\Omega} - \Omega)\, \sqrt{\overline{\Omega}-\Omega}\, \frac{(\underset{\sim}{p}^2+4m^2)^{5/4}}{2m} \;\; F(\Omega\, ,\hat{\underset{\sim}{P}},P^-,Q,U,V) \qquad (7.15)$$

with F the angular integral over f' . This is again in $H_\varepsilon(\ldots;V)$** . For $\underset{\sim}{p}=0$ we recover the result (7.10). (The functions F are not the same in (7.10) and (7.15)!) We have treated the case $\underset{\sim}{p}=0$ separately, not out of necessity, but simply as an example where the emergence of a singularity in $\Omega=\overline{\Omega}$ can be seen very explicitly.

The remaining $\hat{\underset{\sim}{P}}$-integrations in (7.6) do not destroy the H_ε character of the integrand, as can be shown by estimates of the type used in the 1st case.

We proceed to the <u>higher orders</u> $\sigma > 1$. We assume the theorem to be true in all orders $\tau<\sigma$ and wish to prove it in order σ .

First we prove existence of the integrals

$$J(K,P,Q) = \int dW \prod_1^{\ell} \delta_+(w_i)\, \tilde{g}^+_{\mu,\tau}(K_L,P_L,Q_L;-W)\, \tilde{g}^+_{\nu\sigma-\tau}(K_R,P_R,Q_R;W) \;, \qquad (7.16)$$

$1\leqslant\tau\leqslant\sigma-1$, and show that they have the property claimed for $\tilde{g}_{\mu\sigma}$ in Theorem 7.1 . As usual: $K_L \vee K_R = K$, $K_L \wedge K_R = \emptyset$, etc. Define $f(K,..,V)$ as in Theorem 7.1 . Consider

* A similar discussion is carried through in more detail in Ref.[31], Section 3.

**The case $\underset{\sim}{p}\neq 0$ could also be reduced to the simpler case $\underset{\sim}{p}=0$ by considerations of invariance: see Ref.[21].

$$J_1(K_R, \underset{\sim}{P}_R, P^-, Q, U, V, W) = \prod_{Q_L} \chi(-q_i) \prod_1^{\ell} \chi(w_i)$$

$$\times \int dK_L\, d\underset{\sim}{P}_L \prod_{P_L} \chi(p_i)\, \tilde{g}^+_{\mu,\tau}(K_L, P_L, Q_L; -W)\, f(K, \underset{\sim}{P}, P^-, Q, U, V) \quad . \tag{7.17}$$

By the induction hypothesis J_1 exists and is in $H_{\varepsilon''}(K_R, P_R, P^-, Q_L, U, W; Q_R, V)$, $\varepsilon'' < \varepsilon$. Next we form

$$J_2(W^-, P^-, Q, U, V) = \prod_{Q_R} \chi(-q_i) \int dK_R\, d\underset{\sim}{P}_R \prod_{P_R} \chi(p_i)\, d\underset{\sim}{W} \prod_1^{\ell} \frac{\chi(w_i)}{2\omega(\underset{\sim}{w}_i)}$$

$$\times \tilde{g}^+_{\mu,\sigma-\tau}(K_R, P_R, Q_R; W)\, J_1(K_R, \underset{\sim}{P}_R, W, P^-, Q, U, V) \quad . \tag{7.18}$$

The factor $\Pi[2\omega(\underset{\sim}{w}_i)]^{-1}$ can be multiplied into J_1 without changing its properties. We obtain then from the induction hypothesis, that J_2 is in $H_{\varepsilon'}(W^-, P^-, Q, U; V)$, $\varepsilon' < \varepsilon''$. This implies in particular that the restriction of J_2 to the mass shell $w_i^- = 0$ exists and is in $H_{\varepsilon'}(P^-, Q, U; V)$:

$$J_3(P^-, Q, U, V) = \int dW^- \prod_1^{\ell} \delta(w_i^-)\, J_2(W^-, P^-, Q, U, V)$$

$$= \int dK\, d\underset{\sim}{P}\, J(K, P, Q)\, f(K, P, Q, U, V) \tag{7.19}$$

exists and is in $H_{\varepsilon'}(P^-, Q, U; V)$. Note that this property of J implies that J is a tempered distribution.

We have now proved that the right-hand sides of the completeness equations (3.9) and (6.39) exist as tempered distributions, so that the construction of r_σ , $g_{\mu,\sigma}$ explained in Chapters III-VI makes sense: r_σ and $g_{\mu,\sigma}$ exist as tempered distributions. We must still prove that they satisfy theorem 7.1 . Because $\tilde{g}_{\mu,\sigma}(P)$ is equal to $\tilde{r}_\sigma(P)$ plus a sum of terms of the form (7.16) it suffices to prove the theorem for $\tilde{r}_\sigma$.

We start from the expression (4.58) for r_σ , adapting it somewhat to our present purposes. Define

$$\zeta_{i\kappa\omega}(x_1,..,x_n) = \zeta_{i\kappa}(x_1^0,\omega\underset{\sim}{x}_1,...,x_n^0,\omega\underset{\sim}{x}_n) \tag{7.20}$$

for $0<\omega\leq 1$. Going through the arguments of Chapter IV we see that we can effect the replacement $\underset{\sim}{x}_i \to \omega\underset{\sim}{x}_i$ in the auxiliary functions ϑ , f_i , ρ_κ without destroying any of their relevant properties. This means that (4.58) remains valid if $\zeta_{i\kappa}$ is replaced by $\zeta_{i\kappa\omega}$. The coefficients $R^D_{\sigma\kappa}$ become, of course, also dependent on ω : $R^D_{\sigma\kappa} \to R^D_{\sigma\kappa\omega}$. In the limit $\omega \to 0$, $\zeta_{i\kappa\omega}$ becomes a function

$$\eta_{i\kappa}(x_1^0,..,x_n^0) = \zeta_{i\kappa}(x_1^0,\underset{\sim}{0},...,x_n^0,\underset{\sim}{0}) \tag{7.21}$$

depending only on the time components of the x_i . $\eta_{i\kappa}$ is in $\mathcal{O}_M$ (C^∞ and polynomially bounded), hence the product $\eta_{i\kappa}\, I_{i\sigma}$ exists. Since the $\lim_{\omega\to 0}$ of the first term on the right of (4.58) exists and the left-hand side is independent of ω , the limit $\bar{R}^D_{\sigma\kappa} = \lim\limits_{\omega\to 0} R^D_{\sigma\kappa\omega}$ exists also. We obtain

$$r_\sigma(x_1,..,x_n) = \lim_{\kappa\to\infty} \{ \sum_{i=2}^{n} \eta_{i\kappa}(x_1^0,..,x_n^0)\, I_{i\sigma}(x_1,..,x_n) - \sum_{|D|\leq N} \bar{R}^D_{\sigma\kappa}\, D \prod_2^n \delta^4(x_1-x_i) \} \ , \tag{7.22}$$

or in p-space

$$\tilde{r}_\sigma(p_1,..,p_n) = \lim_{\kappa\to\infty} \{ \sum_{i=2}^{n} \tilde{r}_{i\sigma\kappa}(P) - \delta^4(\sum_1^n p_i) \sum_D \bar{R}^D_{\sigma\kappa}\, P_D(P) \} \tag{7.23}$$

with P_D a form of degree $|D|$, and

$$\tilde{r}_{i\sigma\kappa}(p_1,..p_n) = \int dq_{1o} \dots dq_{no}\, \tilde{I}_{i\sigma}(p_{1o}-q_{1o},\underset{\sim}{p}_1,...,p_{no}-q_{no},\underset{\sim}{p}_n)\tilde{\eta}_{i\kappa}(q_{1o},..q_{no}) \ , \tag{7.24}$$

$$\tilde{\eta}_{i\kappa}(Q_o) = \int dX^o \exp[i\sum_j^n x_j^o q_{jo}]\, \eta_{i\kappa}(X^o) \quad .$$

$\eta_{i\kappa}$ can be written as a function of the difference variables $\xi_i = x_1 - x_i$, $i = 2,..,n$, only:

$$\eta_{i\kappa}(x_1^o,..,x_n^o) = \eta'_{i\kappa}(\xi_2^o,..,\xi_n^o) \quad . \tag{7.25}$$

Because of the support of $[1-\rho_\kappa]$ (see (4.53)) there exists a compact neighbourhood of the origin $\xi_2=..=\xi_n=0$ outside of which $\eta'_{i\kappa}$ is independent of κ . Together with the support properties of $I_{i\sigma}$ this tells us that, for κ fixed at an arbitrary finite value, the difference

$$\Delta(\xi_2,..,\xi_n) = \Delta(x_1,..,x_n) = r_\sigma(X) - \sum_i r_{i\sigma\kappa}(X) \tag{7.26}$$

has compact support in the variables ξ_i . $\Delta(\Xi)$ can then be written in the form (see [19], Theorem 26, p.91)

$$\Delta(\Xi) = \sum_D D\, F_D(\Xi) \quad , \tag{7.27}$$

where the D are finitely many differential operators and the F_D are continuous functions with compact support. Hence the Fourier transform of $\Delta(X)$ is of the form

$$\tilde{\Delta}(p_1,..,p_n) = \delta^4(\sum_i^n p_i)\, G(p_2,..,p_n) \tag{7.28}$$

with G an entire function which is polynomially bounded for real values of the arguments p_i . $\tilde{\Delta}$ is thus essentially of the form (7.3) . The same proof as used there shows that $\tilde{\Delta}$ satisfies Theorem 7.1 . Therefore the theorem holds for $\tilde{r}_\sigma$ if it holds for $\tilde{r}_{i\sigma\kappa}$, We note that for the same reason the λ-limit introduced in (5.35) does not invalidate our proofs since it affects only terms with compact support in Ξ .

For the discussion of $\tilde{r}_{i\sigma\kappa}$ we need some information on the singularities of $\tilde{\eta}'_{i\kappa}$. Define

$$\eta'_i(\xi^o_2,..,\xi^o_n) = \eta_i(x^o_1,..,x^o_n) = \zeta_i(x^o_1,\underset{\sim}{0},..,x^o_n,\underset{\sim}{0}) \quad . \tag{7.29}$$

The difference $\eta'_i(\Xi^o) - \eta'_{i\kappa}(\Xi^o)$ is of compact support. From this and the scale invariance of η'_i we obtain

$$D\eta'_{i\kappa}(\Xi^o) = \mathcal{O}(|\Xi^o|^{-|D|}) \tag{7.30}$$

for $|\Xi^o|^2 = \sum_2^n (\xi^o_i)^2 \to \infty$, D a differential operator of order $|D|$. Hence $D\eta'_{i\kappa}$ is integrable if $|D| > n-1$. This implies that

$$\mathfrak{t}_N(Q_o)\ \tilde{\eta}'_{i\kappa}(Q_o) \quad ,$$

with $\hat{Q}_o = \{q_{2o},..,q_{no}\}$, is uniformly bounded and $(N-n)$ times continuously differentiable for any form $\mathfrak{t}_N$ of degree $N \geq n$. In particular, $|\hat{Q}_o|^{2N}\ \tilde{\eta}'_{i\kappa}(\hat{Q}_o)$ is $(2N-n)$ times differentiable, so that $\tilde{\eta}'_{i\kappa}$ is C^∞ everywhere except at the origin, and all derivatives of $\tilde{\eta}'_{i\kappa}$ decrease strongly at infinity.

The difference $\tilde{\eta}'_i - \tilde{\eta}'_{i\kappa}$ is an entire function, hence the singularity of $\tilde{\eta}'_{i\kappa}$ at the origin is the same as that of $\tilde{\eta}'_i$. The scale invariance of η'_i gives in p-space

$$\tilde{\eta}'_i(\lambda\hat{Q}_o) = \lambda^{-n+1}\ \tilde{\eta}'_i(\hat{Q}_o) \tag{7.31}$$

for $0<\lambda<\infty$. We see that $\tilde{\eta}'_i$, and therefore $\tilde{\eta}'_{i\kappa}$, has a singularity of order $n-1$ at the origin. $\mathfrak{t}_{n-1}\ \tilde{\eta}'_{i\kappa}$ is still uniformly bounded, but no longer necessarily continuous in the origin.

We note also that $\tilde{\eta}'_{i\kappa}$, being the Fourier transform of the tempered distribution $\eta'_{i\kappa}$, is a tempered distribution. This means that a well-defined prescription of how to integrate over the singularity in the origin exists. We need not know this prescription explicitly. $\tilde{\eta}'_{i\kappa}$ is, in fact, not only defined on $\mathcal{S}$ but even on H_ε . Let $\tilde{\phi}(\hat{Q}_o) \in H_\varepsilon$, and introduce an arbitrary function $\tilde{\psi}(\hat{Q}_o) \in \mathcal{S}$ with $\tilde{\psi} \equiv 1$ in a neighbourhood of the origin . Then

$$\int d\hat{Q}_o \; \tilde{\eta}'_{i\kappa}(\hat{Q}_o) \; \tilde{\phi}(\hat{Q}_o) = \tilde{\phi}(0) \int d\hat{Q}_o \; \tilde{\eta}'_{i\kappa}(\hat{Q}_o) \; \tilde{\psi}(\hat{Q}_o)$$

$$+ \int d\hat{Q}_o \; \tilde{\eta}'_{i\kappa}(\hat{Q}_o) \left[\tilde{\phi}(\hat{Q}_o) - \tilde{\phi}(0) \; \tilde{\psi}(\hat{Q}_o) \right] \; .$$

The first term on the right exists because of $\tilde{\eta}'_{i\kappa} \in \mathcal{S}$. From the Hölder continuity of $\tilde{\phi}$ we obtain

$$|\tilde{\phi}(\hat{Q}_o) - \tilde{\phi}(0) \; \tilde{\psi}(\hat{Q}_o)| \quad \leq \quad c\,|\hat{Q}_o|^\varepsilon$$

in a neighbourhood of the origin. This reduces the order of the singularity of the integrand in the second term to $n-1-\varepsilon$, and this is integrable.

$\tilde{\eta}_{i\kappa}$ is of the form

$$\tilde{\eta}_{i\kappa}(q_{1o},..,q_{no}) = \delta(\sum_1^n q_{jo}) \; \hat{\eta}(q_{1o},..,q_{no}) \quad . \tag{7.32}$$

$\hat{\eta}$ is defined on the manifold $\Sigma \, q_{jo} = 0$ and is given by

$$\hat{\eta}(q_{1o},..,q_{no}) = 2\pi\tilde{\eta}'_{i\kappa}(-q_{2o},..,-q_{no}) \quad . \tag{7.33}$$

Again, the integral of $\tilde{\eta}_{i\kappa}$ over functions from H_ε is defined.

We can now return to the discussion of $\tilde{r}_{i\sigma\kappa}$. Let $f(K,..,V)$ be the function introduced in Theorem 7.1 .

We must discuss

$$\bar{f}(P^-,Q,U,V) = \prod_Q \chi(-q_i) \int dK\, d\underset{\sim}{P}\, \prod_P \chi(p_i)\, dK_0'\, dP_0'\, dQ_0' \quad \tilde{\eta}_{i\kappa}(K_0-K_0',P_0-P_0',Q_0-Q_0')$$

$$\times \quad \tilde{I}_{i\sigma}(K_0',\underset{\sim}{K},P_0',\underset{\sim}{P},Q_0',\underset{\sim}{Q})\ f(K;P,Q,U,V) \quad , \qquad (7.34)$$

where $p_{jo} = p_j^- + \omega(\underset{\sim}{p}_j)$ has to be substituted everywhere. If we introduce $p_i'^- = p_{io}' - \omega(\underset{\sim}{p}_i)$ as new integration variables instead of p_{io}' , then (7.34) becomes

$$\bar{f}(P^-,..,V) = \Pi\chi(-q_i) \int dK\, d\underset{\sim}{P}\, \Pi\, \chi(p_i)\, dK_0'\, dP'^-\, dQ_0'$$

$$\times\ \tilde{\eta}_{i\kappa}(K_0-K_0',P^- - P'^-,Q_0-Q_0')\ \tilde{I}_{i\sigma}(K_0',\underset{\sim}{K},P'^-,\underset{\sim}{P},Q_0',\underset{\sim}{Q})\ f(K,\underset{\sim}{P},P^-,Q,U,V) \quad .$$

$$(7.35)$$

Let $\chi^0(p)$ be a function with the properties of $\chi(p)$ (see Chapter II) but with a wider support - still satisfying (2.41) - such that $\chi^0(p) \equiv 1$ in an open neighbourhood of supp $\chi(p)$. We define $\bar{f}_1(P^-,Q,U,V)$ by (7.35) with the additional factor

$$\Phi(P'^-,\underset{\sim}{P},Q_0',\underset{\sim}{Q}) = \prod_1^\gamma \chi^0(-q_{jo}',-\underset{\sim}{q}_j)\ \prod_1^\beta \chi^0(p_j'^- + \omega(\underset{\sim}{p}_j),\underset{\sim}{p}_j) \qquad (7.36)$$

multiplied into the integrand. $\bar{f}_2(P^-,Q,U,V)$ is defined by

$$\bar{f} = \bar{f}_1 + \bar{f}_2 \quad . \qquad (7.37)$$

We show that $\bar{f}_1$, $\bar{f}_2$ both possess the required Hölder property. We start with the simpler case, that of $\bar{f}_2$. The product

$$\Pi\ \chi(-q_j)\ \Pi\ \chi(p_j)\ [1 - \Phi(P'^-,\underset{\sim}{P},Q_0',\underset{\sim}{Q})] \qquad (7.38)$$

vanishes in a neighbourhood of $P'^- = P^-$, $Q'_o = Q_o$. This is the case, in particular, in the singularity of $\hat{\eta}$, which is therefore irrelevant: without changing the value of $\bar{f}_2$ we can replace $\hat{\eta}$ by a C^∞-function $\hat{\eta}^o$ which coincides with $\hat{\eta}$ in the support of (7.38). $\hat{\eta}^o$ decreases strongly in K'_o , P'^- , Q'_o for K_o , Q_o , P^- fixed. Hence

$$\int dK'_o \, dP'^- \, dQ'_o \;\; \tilde{\eta}^o_{i\kappa}(K_o-K'_o, P^--P'^-, Q_o-Q'_o) \; \tilde{I}_{i\sigma}(K'_o, \underset{\sim}{K}, P'^-, \underset{\sim}{P}, Q'_o, \underset{\sim}{Q})$$

$$\times \left[1 - \Phi(P'^-, \underset{\sim}{P}, Q'_o, \underset{\sim}{Q})\right] = \delta^4(\Sigma k_j + \Sigma p_j + \Sigma q_j) \; g(K,P,Q)$$

with g a polynomially bounded C^∞ function. $\tilde{\eta}^o_{i\kappa}$ is defined from $\hat{\eta}^o$ in analogy to (7.32). The remaining integral

$$\bar{f}_2 = \Pi \, \chi(-q_j) \int dK \, d\underset{\sim}{P} \, \Pi\chi(p_j) \; \delta^4(\Sigma k_j + \Sigma p_j + \Sigma q_j)$$

$$\times \; g(K,\underset{\sim}{P},P^-,Q) \; f(K,\underset{\sim}{P},P^-,Q,U,V)$$

exists and has the desired properties, as can be shown by the arguments used in the case $\sigma = 1$: gf has the same properties as f alone.

We come to $\bar{f}_1$. After the change of variables $K_o \to K'_o - K_o$, $\bar{f}_1$ reads

$$\bar{f}_1 = \Pi \, \chi(-q_i) \int dK_o \, d\underset{\sim}{K} \, dK'_o \, d\underset{\sim}{P} \, dP'^- \, dQ'_o \; \Pi \, \chi(p_i) \; \Pi\chi^o(p'^-_i + \omega(\underset{\sim}{p}_i), \underset{\sim}{p}_i)$$

$$\times \; \Pi \, \chi^o(-q'_{io}, -\underset{\sim}{q}_i) \; \tilde{\eta}_{i\kappa}(-K_o, P^--P'^-, Q_o-Q'_o) \; \tilde{I}_{i\sigma}(K'_o, \underset{\sim}{K}, P'^-, \underset{\sim}{P}, Q'_o, \underset{\sim}{Q})$$

$$\times \quad f(K'_o-K_o, \underset{\sim}{K}, \underset{\sim}{P}, P^-, Q, U, V) \; . \qquad (7.39)$$

We have

$$\prod \chi(p_i)\ f(K_o^{\prime}-K_o,\underset{\sim}{K},P,Q,U,V) \in H_\varepsilon(K_o^{\prime},\underset{\sim}{K},P,U;Q,K_o,P^{\prime -},Q_o^{\prime},V) \quad .$$

Hence, according to an earlier result proving the existence of $\tilde{I}_{i\sigma}$, the partial integral

$$J_1(K_o,P^{\prime -},Q_o^{\prime},P^-,Q,U,V) = \prod_i \chi^o(-q_{io}^{\prime},-\underset{\sim}{q}_i) \int dK_o^{\prime}\ d\underset{\sim}{K}\ d\underset{\sim}{P}\ \prod_i \chi(p_i)$$

$$\times \prod_i \chi^o(p_i^{\prime -}+\omega(\underset{\sim}{p}_i),\underset{\sim}{p}_i)\ \tilde{I}_{i\sigma}(K_o^{\prime},\underset{\sim}{K},P^{\prime -},\underset{\sim}{P},Q_o^{\prime},\underset{\sim}{Q})\ f(K_o^{\prime}-K_o,\underset{\sim}{K},P,Q,U,V)$$

is in $H_\varepsilon(P^{\prime -},Q_o^{\prime},P^-,\underset{\sim}{Q},U;K_o,Q_o,V)$. Multiplication of J_1 with $\prod \chi(-q_i)$ gives J_2 , with

$$J_2(K_o,\dots,V) \in H_\varepsilon(P^{\prime -},Q_o^{\prime},P^-,Q,U;K_o,V) \quad .$$

This leaves us with

$$\bar{f}_1 = \int dK_o\ dP^{\prime -}\ dQ_o^{\prime}\ \tilde{\eta}_{i\kappa}(-K_o,P^--P^{\prime -},Q_o-Q_o^{\prime})\ J_2(K_o,P^{\prime -},Q_o^{\prime},P^-,Q,U,V) \quad . \tag{7.40}$$

We note that for V fixed the integrand decreases strongly in all directions. The behaviour at infinity will thus cause no difficulties, so that we can concentrate on proving the Hölder continuity of $\bar{f}_1$ *.

We have already noted that the distribution $\tilde{\eta}_{i\kappa}$ is defined on Hölder continuous test functions, so that $\bar{f}_1$ is defined for all values of its arguments.

---

* We shall henceforth dispense with discussing the behaviour at infinity explicitly every time we ought to do so: it can be done easily and gives always the desired result.

We use the δ-factor in $\tilde{\eta}_{i\kappa}$ to perform one of the integrations: the k_{1o}-integration if $\alpha>0$, the q'_{1o}-integration if $\alpha=0$, $\beta,\gamma \geq 2$. In either case we obtain an expression of the following structure:

$$\bar{F}(x_1,\ldots,x_n,y_1,\ldots,y_m) = \int dz_1..dz_n\, dw_1..dw_s\, H(x_1-z_1,..,x_n-z_n,w_1,..,w_s)$$

$$\times\; F(x_1,..,x_n,z_1,..,z_n,w_1,..,w_s,y_1,..,y_m)\ . \quad (7.41)$$

The x_i ,... are the original variables $p_{\bar{i}}$, $q_{i\mu}$, etc., suitably renamed. H is the function $\tilde{\eta}'_{i\kappa}$ written in the appropriate variables, with $\tilde{\eta}'_{i\kappa}$ having the properties discussed above. F is Hölder continuous of index ε in all variables and decreases strongly in the relevant directions, so that the integral (7.41) exists at infinity. We wish to prove that $\bar{F}$ is Hölder continuous of index $\varepsilon'<\varepsilon$. The proof is a generalisation of the proof of the Plemelj-Privalov theorem, as given in Ref.[22], sections 9 and 10 .

Let $\psi(X,W)\in\mathcal{D}$ (i.e. C^∞ and with compact support), with $\psi\equiv 1$ in a neighbourhood of the origin. We split $\bar{F}$ as follows:

$$\bar{F} = \bar{F}_1 + \bar{F}_2\ , \quad (7.42)$$

$$\bar{F}_1(X,Y) = \int dZ\, dW\, H(X-Z,W)\, \psi(X-Z,W)\, \hat{F}(X,Y)\ , \quad (7.43)$$

$$\bar{F}_2(X,Y) = \int dZ\, dW\, H(X-Z,W)\, G(X,Z,W,Y)\ , \quad (7.44)$$

with

$$\hat{F}(X,Y) = F(X,Z,W,Y)\Big|_{Z=X,W=0}\ , \quad (7.45)$$

$$G(X,Z,W,Y) = F(X,Z,W,Y) - \psi(X-Z,W)\,\hat{F}(X,Y)\ . \quad (7.46)$$

In $\bar{F}_1$ we introduce $x_i - z_i$ as integration variables instead of z_i :

$$\bar{F}_1(X,Y) = \hat{F}(X,Y) \int dZ\, dW\, H(Z,W)\, \psi(Z,W) \quad .$$

The integral exists and is independent of X , Y . $\hat{F}(X,Y)$ is clearly Hölder continuous of index ε , hence the same is true for $\bar{F}_1$.

We turn to $\bar{F}_2$ and consider

$$\begin{aligned} & D(X,Y,\Delta X,\Delta Y) \\ & = \bar{F}_2(X+\Delta X,Y+\Delta Y) - \bar{F}_2(X,Y) \\ & = \int dZ\, dW\, \{H(X-Z+\Delta X,W)\, G(X+\Delta X,Z,W,Y+\Delta Y) - H(X-Z,W)\, G(X,Z,W,Y)\} \quad . \end{aligned} \tag{7.47}$$

Introduce $A^2 = |\Delta X|^2 + |\Delta Y|^2$. We split D into two parts: $D = D_1 + D_2$, where D_1 is the integral (7.47) taken over the region $R^2 = |X-Z|^2 + |W|^2 \leq 4A^2$, D_2 the integral over $R^2 > 4A^2$. Consider D_1 first:

$$D_1 = - D_{11} + D_{12}$$

with

$$D_{11}(X;Y) = \int_{R \leq 2A} dZ\, dW\, H(X-Z,W)\, G(X,Z,W,Y) \quad ,$$

D_{12} the same with X, Y replaced by $X+\Delta X$, $Y+\Delta Y$ in the integrand but not in the region of integration. Because of $G(X,X,0;Y) = 0$ and the Hölder continuity of G we have

$$|G(X,Z,W,Y)| \leq M(X,Z,W,Y)\, (|X-Z|^2 + |W|^2)^{\varepsilon/2} \tag{7.48}$$

with M positive, continuous, and well-behaved at infinity. H has for X = Z , W = 0 a singularity of order n+s :

$$H = (|X-Z|^2 + |W|^2)^{-\frac{n+s}{2}} \hat{H}(X-Z,W) \qquad (7.49)$$

with $\hat{H}$ bounded everywhere and continuous outside the origin. This yields

$$|D_{11}| \leqslant \int_{R\leqslant 2A} dZ\ dW\ |\hat{H}|\ M\ (|X-Z|^2 + |W|^2)^{\frac{n+s-\varepsilon}{2}} \leqslant C(X,Y)\ A^{\varepsilon}$$

for $A \to 0$, with C positive and continuous. A similar estimate can be made for D_{12} , so that D_1 satisfies the desired inequality with index ε .

We split D_2 differently:

$$D_2 = D_{21} + D_{22} \ ,$$

$$D_{21} = \int_{R\geqslant 2A} dZ\ dW\ H(X-Z,W)\ \{G(X+\Delta X,Z,W,Y+\Delta Y) - G(X,Z,W,Y)\}\ .$$

$$D_{22} = \int_{R\geqslant 2A} dZ\ dW \{H(X-Z+\Delta X,W) - H(X-Z,W)\}\ G(X+\Delta X,Z,W,Y+\Delta Y)\ .$$

For D_{21} we use (7.49) and

$$|G(X+\Delta X,Z,W,Y+\Delta Y) - G(X,Z,W,Y)| \leqslant A^{\varepsilon}\ M'(X,Z,W,Y)\ , \qquad (7.50)$$

M' continuous, and obtain

$$|D_{21}| \leqslant A^{\varepsilon} \int_{R\geqslant 2A} dZ\ dW \frac{|\hat{H}|\ M'}{(|X-Z|^2 + |W|^2)^{\frac{n+s}{2}}} \leqslant C'(X,Y)\ A^{\varepsilon} |\log A|\ ,$$

the integral diverging logarithmically for $A \to 0$. C' is continuous. But $A^{\varepsilon}|\log A|$ is majorised for small A by $A^{\varepsilon'}$ for any $\varepsilon' < \varepsilon$, hence the D_{21} part gives Hölder continuity of index ε' in $\overline{F}_2$.

In D_{22} we note that for Z, W in the domain of integration, H is C^∞ in the interval $[(X-Z+\Delta X,W)\ ,\ (X-Z,W)]$. We apply the mean value theorem:

$$H(X-Z+\Delta X,W) - H(X-Z,W) = \Delta X \cdot \mathrm{grad}_X\, H(\bar{X}-Z,W) \quad ,$$

with $\bar{X}$ lying between X and $X+\Delta X$. $\partial_{x_i} H(X-Z,W)$ is C^∞ everywhere except in $X=Z$, $W=0$ where it has a singularity of order $n+s+1$:

$$\partial_{x_i} H(X-Z,W) = R^{-n-s-1}\, \hat{H}_i(X-Z,W) \tag{7.51}$$

with $\hat{H}_i$ bounded everywhere and continuous everywhere except for $X=Z$, $W=0$. In the region of integration we have

$$|\bar{X}-Z| \geq |X-Z| - |X-\bar{X}| \geq |X-Z| - A \geq |X-Z| - \tfrac{1}{2}|X-Z| = \tfrac{1}{2}|X-Z| \ ,$$

with which we obtain from (7.51)

$$|\partial_{x_i} H(\bar{X}-Z,W)| < 2^{n+s}\, R^{-n-s-1}\, |\hat{H}_i(\bar{X}-Z,W)| \quad .$$

This together with (7.48) yields

$$|D_{22}| \leq A\, 2^{n+s} \int_{R\geq 2A} dZ\, dW\, R^{-n-s-1+\varepsilon} \sum_i |H_i(X-Z,W)|\, M(X,Z,W,Y)$$

$$\leq C''(X,Y)\, A^{\varepsilon}$$

for $A \to 0$, with C'' continuous.

This completes the proof of Theorem 7.1 . It has already been noted in the course of this proof that the theorem implies the existence of the integrals (3.11) and (6.39). Furthermore, property iii) of Theorem 2.1 is clearly a corollary of Theorem 7.1 . It remains to be seen that $\int dX\, G_{\mu,\sigma}(X) \quad \phi(X)$, $\phi \in \mathcal{f}$, as defined by (6.40) is defined on $\mathcal{L}_{\varepsilon}^{in}$ and maps this domain into $\mathcal{L}_{\varepsilon'}^{in}$.

Let

$$\Phi = \int \prod_{i=1}^{a} \frac{d\underset{\sim}{p}_i}{2\omega(\underset{\sim}{p}_i)} \quad \hat{f}(\underset{\sim}{P})\, \hat{A}^{in*}(\underset{\sim}{p}_1)\ldots\hat{A}^{in*}(\underset{\sim}{p}_a)\; \Omega \tag{7.52}$$

with $\hat{f} \in H_\varepsilon$.

Equation(6.40) reads in p-space

$$\int dK\, \tilde{G}_{\mu\sigma}(K)\, \tilde{\phi}(K) = \sum_{\ell=0}^{\infty} \frac{(2\pi)^{\ell}}{\ell!} \int dK\, dP\, \tilde{\phi}(K)\, \tilde{g}^{+}_{\mu\sigma}(K;-P) \quad :\tilde{A}^{in}(p_1)\ldots\tilde{A}^{in}(p\ ): \tag{7.53}$$

The series breaks off after a finite number of terms. It suffices therefore to consider an individual term. The ℓth term applied to Φ gives a vector with components with $a-\ell, a-\ell+2,\ldots,a+\ell$ in-particles. The corresponding wave functions are sums of terms of the form

$$\hat{g}(\underset{\sim}{q}_1,\ldots,\underset{\sim}{q}_\alpha,\underset{\sim}{q}'_1,\ldots\underset{\sim}{q}'_\beta) = \text{const.} \int dK\, dP\, dQ^- \prod_{1}^{\gamma} \delta_+(p_j) \prod_{1}^{\alpha} \delta(q_j^-)$$

$$\times\ \tilde{g}^{+}_{\mu\sigma}(K;-P,Q)\, \tilde{\phi}(K)\quad \hat{f}(\underset{\sim}{P},\underset{\sim}{Q}') \quad . \tag{7.54}$$

Obviously $\tilde{\phi}\hat{f} \in H_\varepsilon(K,\underset{\sim}{P},\underset{\sim}{Q}';P^-,Q)$, from which we derive with the help of Theorem 7.1 that $\hat{g} \in H_{\varepsilon'}$, $\varepsilon' < \varepsilon$, q.e.d.

In the case of the expansion (6.41) for the field $\tilde{A}_\sigma(k)$ the integrands in (7.53) contain also the de-amputation factor $(k^2-m^2-ik_o\varepsilon)^{-1}$. We can write it

$$\frac{1}{k^2-m^2-ik_o\varepsilon} = \frac{\chi(k)}{k^2-m^2-i\varepsilon} + \frac{\chi(-k)}{k^2-m^2+i\varepsilon} + \text{ a } C^\infty\text{-function.} \tag{7.55}$$

The C^∞-part obviously does not cause any trouble. The other two terms are concentrated in a neighbourhood of the mass shell. Let us consider, for example, the first term and its contribution to an integral of the type (7.54) . Integration over all variables except k^- gives a Hölder continuous function in k^- , because of Theorem 7.1 , and the integral of such a function over $(k^- - i\varepsilon)^{-1}$ exists. An analogous argument holds for the second term.

VIII. RENORMALISABLE THEORIES

We return to the discussion of the ambiguities that are still left after application of the local regularity requirement introduced in Chapter V . It is obviously interesting to know exactly how ambiguous a given theory, i.e. a theory of given type (μ,ν) is*.

According to Chapter V the number of ambiguities in $r_\sigma(x_1,..x_n)$ depends on its s-degree $SD(r_\sigma) = d(n,\sigma)$. These s-degrees will now be studied.

A theory of type (μ,ν) is characterised by the vanishing of all first order functions $r_1(x_1,..x_n)$ except the μ-point function. Therefore

$$d(n,1) = \infty \quad \text{for } n \neq \mu \quad . \tag{8.1}$$

$\tilde{r}_1(p_1,..p_\mu)$ is of the form

$$\tilde{r}_1(p_1,..,p_\mu) = \delta^4(\Sigma\, p_i)\mathcal{P}(P) \quad , \tag{8.2}$$

---

* See Chapter IV, Eq.(4.6) for the definition of the type.

P a form of degree ν , so that

$$\tilde{r}_1(\lambda P) = \lambda^{-4+\nu} \tilde{r}_1(P) \quad . \tag{8.3}$$

Using the p-space definition (5.5) we obtain

$$d(\mu,1) = -4(\mu-1) - \nu \quad . \tag{8.4}$$

For the discussion of the higher orders it is useful to know that the behaviour of $\tilde{r}_\sigma(\lambda p_1,.,\lambda p_n)$ for $\lambda\to\infty$ is closely related to the behaviour of $\tilde{r}_\sigma(p_1,..p_n)$ for $m\to 0$, i.e. with the infrared behaviour. This is so because of

<u>Lemma 8.1</u>

Let $r(x_1,..,x_n)$ be distributions satisfying the assumptions of Theorem 2.2 . Then

$$\underline{r}^\lambda(x_1,..,x_n) = \lambda^{-3n} r(\lambda^{-1}x_1,..,\lambda^{-1}x_n) \quad , \quad 0 < \lambda < \infty \quad , \tag{8.5}$$

satisfy again the assumptions of Theorem 2.2 , but with the mass $m\lambda^{-1}$ instead of m .

Eq. (8.5) reads in p-space

$$\tilde{\underline{r}}^\lambda(p_1,..,p_n) = \lambda^n \tilde{r}(\lambda p_1,..,\lambda p_n) \quad . \tag{8.6}$$

It is obvious that $\underline{r}^\lambda$ satisfies conditions i), ii) and iii) of Theorem 2.1 , with condition iii) of course formulated for the new mass shell $p^2 = m^2/\lambda^2$. In order to see that $\underline{r}^\lambda$ solves the completeness equations with mass m/λ we use the p-space form (2.35) of these equations and note

$$\delta_+(\lambda k;m) = \theta(k_o)\ \delta(\lambda^2 k^2 - m^2) = \lambda^{-2}\theta(k_o)\ \delta\left(k^2 - \frac{m^2}{\lambda^2}\right) = \lambda^{-2}\delta_+(k;m/\lambda) \quad . \tag{8.7}$$

The rest is a trivial calculation.

The normalisation condition (2.58) remains satisfied under the transformation (8.6):

$$\tilde{\underline{r}}^\lambda(p,q) = \lambda^2\ \delta^4(\lambda p+\lambda q)\ \{\ \frac{1}{2\pi}\ (\lambda^2 q^2 - m^2) + (\lambda^2 q^2 - m^2)^2\ F(\lambda q)\}$$

$$= \delta^4(p+q)\ \{\ \frac{1}{2\pi}\left(q^2 - \frac{m^2}{\lambda^2}\right) + \left(q^2 - \frac{m^2}{\lambda^2}\right)^2\ F^\lambda(q)\} \tag{8.8}$$

with $F^\lambda(q) = \lambda^2\ F(\lambda q)$ analytic in $q^2 < 4m^2/\lambda^2$.

We apply this lemma to our theory of type (μ,ν) . From (8.2) we obtain

$$\tilde{\underline{r}}_1^\lambda(p_1,..,p_\mu) = \lambda^{\nu+\mu-4}\ \delta^4(\Sigma\ p_i)\ \mathcal{P}(P) \quad . \tag{8.9}$$

We remember that in the perturbative series (3.1) for r , r_1 occurs multiplied with the coupling constant g . We define

$$g_\lambda = g\lambda^{\nu+\mu-4} \quad , \tag{8.10}$$

so that

$$g\ \tilde{\underline{r}}_1^\lambda(p_1,..,p_\mu) = g_\lambda\ \delta^4(\Sigma\ p_i)\ \mathcal{P}(p_1,..,p_\mu) = g_\lambda\ \tilde{r}_1(p_1,..,p_\mu)\ . \tag{8.11}$$

We can carry out the constructions of the preceding chapters for the mass m/λ instead of m and the expansion parameter g_λ instead of g, but with the same $\tilde{r}_1$ given by (8.2). The resulting $\tilde{r}_\sigma$ we call $\tilde{r}^\lambda_\sigma(P)$. In the same way we define $\tilde{I}^\lambda_\sigma$, $\tilde{J}^\lambda_{\ell L\sigma}$, etc.
By definition:

$$\tilde{r}^\lambda_1(P) = \tilde{r}_1(P) \tag{8.12}$$

independent of λ. From (8.10) and Lemma 8.1 we obtain

$$\tilde{r}^\lambda_\sigma(p_1,..,p_n) = \lambda^{n-\sigma(\mu+\nu-4)}\ \tilde{r}_\sigma(\lambda p_1,..,\lambda p_n) \tag{8.13}$$

as a possible solution in the higher orders.

In order to determine the s-degree of r_σ we must find the behaviour of $\lambda^{4n}\ \tilde{r}_\sigma(\lambda P)$ for large λ. Because of (8.13) this is equivalent to looking at the $\lambda\to\infty$ limit of

$$\lambda^{3n+\sigma(\mu+\nu-4)}\ \tilde{r}^\lambda_\sigma(p_1,..,p_n) \quad .$$

We see that

$$d(n,\sigma) \geq -3n-\sigma(\mu+\nu-4) \quad , \tag{8.14}$$

provided that $\lambda^{-\varepsilon}\tilde{r}^\lambda_\sigma(P)$ remains bounded for $\lambda\to\infty$ for each $\varepsilon>0$. If $\tilde{r}^\lambda_\sigma$ diverges for $\lambda\to\infty$ like a power of λ or stronger, then the estimate (8.14) must be altered accordingly.

The question of convergence or divergence of $\lim\limits_{\lambda\to\infty} \lambda^{-\varepsilon}\tilde{r}^\lambda_\sigma$ is clearly the question for the infrared divergences occurring when the mass m/λ tends to 0. This problem will be studied next. We want to show that

$$\lim_{\lambda\to\infty} \lambda^{-\varepsilon}\, \tilde{r}_\sigma^\lambda(p_1,\ldots,p_n) = 0 \qquad (8.15)$$

in $\mathcal{S}'$ for any $\varepsilon>0$. For doing this we must refine the estimates of Chapter VII to some extent. We use the definitions and notations introduced there.

We say that a family of functions $f_\lambda(U,V)\in H_\varepsilon(U;V)$, $1 \leq \lambda < \infty$, is <u>bounded</u>, if

$$S_N(f_\lambda|V) \leq S_N(V) \qquad (8.16)$$

uniformly in λ for all N , with S_N polynomially bounded in V . In our application of this notion some of the variables u_i are of the form

$$p_{i\lambda}^- = p_{io} - \omega_\lambda(\underset{\sim}{p}_i) , \quad \omega_\lambda(\underset{\sim}{p}_i) = \sqrt{\underset{\sim}{p}_i^{\,2} + \frac{m^2}{\lambda^2}} , \qquad (8.17)$$

i.e. they depend themselves on the parameter λ . However,

$$\lim_{\lambda\to\infty} p_{i\lambda}^- = p_{io} - |\underset{\sim}{p}_i|$$

exists and is Hölder continuous of index 1 , so that this λ-dependence of u_i does not create any problems.

Let $f_\lambda(K,P,Q,U,V)\in H_\varepsilon(K,\underset{\sim}{P},U;P_\lambda^-,Q,V)$ be such a bounded family of functions of the type considered in Theorem 7.1 . We claim:

<u>Theorem 8.2</u>

Let $\tilde{r}_\sigma^\lambda(K,\underset{\sim}{P},P_\lambda^-,Q)$, $f_\lambda(K,P,Q,U,V)$ be defined as above and

$$\bar{f}_\lambda(P_\lambda^-,Q,U,V) = \prod_1^\gamma \chi(-\lambda q_j) \int dK\, d\underset{\sim}{P} \prod_1^\beta \frac{\chi(\lambda p_j)}{\omega_\lambda(\underset{\sim}{p}_j)}\, \tilde{r}_\sigma^\lambda(K,\underset{\sim}{P},P_\lambda^-,Q)\, f_\lambda(K,\underset{\sim}{P},P_\lambda^-,Q,U,V). \qquad (8.18)$$

Then there exists a positive constant c_σ independent of ε such that $\lambda^{-c_\sigma \varepsilon}\ \bar{f}_\lambda(P^-_\lambda,..,V)$ is a bounded family in $H_{\varepsilon'}(P^-_\lambda,Q,U;V)$ for $\varepsilon' < \varepsilon$.

The proof proceeds in close analogy to the proof of Theorem 7.1 . We start again by discussing the case $\sigma=1$. $\tilde{r}^\lambda_1$ does not depend on λ by definition. f_λ is bounded in λ . Hence (8.18) is almost of the form discussed in the proof of Theorem 7.1 ..It differs from that form by the additional factors $[\omega_\lambda(\underset{\sim}{p}_i)]^{-1}$, the scaling factor λ in the χ-functions, and the λ-dependence of the variables P^-_λ .

About the χ-factors we note that the diameter of the support of $\chi(\lambda p)$ shrinks like $1/\lambda$ for increasing λ , so that

$$\lambda^{-\varepsilon} \sup_{|a|\leq 1} \frac{|\chi(\lambda(p+a))-\chi(\lambda p)|}{|a|^\varepsilon}$$

remains bounded. Hence the λ-dependence of χ is in agreement with the claim of Theorem 8.2 if we choose $c_\sigma > 1$.

The remaining λ-dependence is more difficult to deal with. Note that $[\omega_\lambda(\underset{\sim}{p})]^{-1}$ becomes singular in $\underset{\sim}{p}=0$ in the limit $\lambda\to\infty$. In the case $\alpha>o$ it is easy to see that the estimates of Chapter VII are not affected by the changes. The $|\underset{\sim}{p}_j|^{-1}$ singularities of the ω_λ^{-1} factors are too weak to destroy the existence of the $\underset{\sim}{p}_j$ -integrals. For $\alpha=0$ we look first at the case $\underset{\sim}{p}\neq 0$. We start from (7.14) which becomes now

$$E = \frac{\lambda}{m}(\underset{\sim}{p}^2+4m^2\lambda^{-2})^{5/4} \int d^3t\ \delta(\Omega_\lambda-\Omega_\lambda+2t^2+ \text{higher terms}) \frac{f''_\lambda(\underset{\sim}{q},\ldots)}{\omega_\lambda(\frac{P}{2}+\underset{\sim}{q})\ \omega_\lambda(\frac{P}{2}-\underset{\sim}{q})} \tag{8.19}$$

where the values (7.13), with m replaced by m/λ , have to be substituted for $\underset{\sim}{q}$. $f''_\lambda(\underset{\sim}{q},\hat{\underset{\sim}{P}},..,V)$ is bounded in $H_\varepsilon(\underset{\sim}{q},\hat{\underset{\sim}{P}},..;V)$* .

---

* We include in f''_λ a factor $\lambda^{-\varepsilon}$ taking care of the χ-functions.

The variables P,..,V will be omitted in the rest of this consideration.

The two ω_λ^{-1} factors can be absorbed into f''_λ . For $\Omega_\lambda \neq \bar{\Omega}_\lambda$ they are not singular enough to influence the estimates. For $\Omega_\lambda = \bar{\Omega}_\lambda$ we integrate only over $q \sim 0$, and there they are not singular at all (for $p \neq 0$).

Define

$$A_\lambda = p^2 + 4m^2\lambda^{-2}$$

If we neglect the higher terms in δ , introduce polar coordinates $t_1 = r\sin\theta\cos\phi$, $t_2 = r\sin\theta\sin\phi$, $t_3 = r\cos\theta$, and integrate over ϕ and r , we obtain

$$E = \frac{\lambda}{m} A_\lambda^{5/4}\sqrt{\Omega_\lambda-\bar{\Omega}_\lambda} \int_{-1}^{1} d(\cos\theta)\, g_\lambda(A_\lambda^{\frac{1}{4}}\sqrt{\Omega_\lambda-\bar{\Omega}_\lambda}\sin\theta \;,\; - A_\lambda^{\frac{3}{4}}\frac{\lambda}{m}\sqrt{\Omega_\lambda-\bar{\Omega}_\lambda}\cos\theta\;) \quad ,$$

$g_\lambda(u,v) \in H_\varepsilon$. With the substitution $\psi = \frac{\lambda}{m} A_\lambda^{\frac{3}{4}}\sqrt{\Omega_\lambda-\bar{\Omega}_\lambda}\cos\theta$ this becomes

$$E = A_\lambda^{\frac{1}{2}} \int_{-\psi_0}^{\psi_0} d\psi\; g_\lambda(m\lambda^{-1} A_\lambda^{-\frac{1}{2}} [A_\lambda^{3/2}\lambda^2 m^{-2} (\Omega_\lambda-\bar{\Omega}_\lambda) - \psi^2]^{\frac{1}{2}} \;,\psi)$$

with $\psi_0 = A_\lambda^{\frac{3}{4}}\lambda m^{-1}\sqrt{\Omega_\lambda-\bar{\Omega}_\lambda}$. Due to the strong decrease of g_λ for $\psi \to \infty$ the integral and its Hölder norms remain bounded for $\lambda \to \infty$. The same is true for the factor $A_\lambda^{\frac{1}{2}}$, i.e. nothing bad happens for $\lambda \to \infty$.

This estimate breaks down for $p \to 0$ because the ω_λ^{-1} singularities coalesce and because the expansion (7.11) is not valid if both p and m are small. In order to see what happens in this limit we carry out the p_2-integration exactly for $p=0$. Instead of (7.10) we obtain

$$(\Omega_\lambda^2 - 4m^2\lambda^{-2}) \frac{\sqrt{\Omega_\lambda^2-4m^2\lambda^{-2}}}{\Omega_\lambda} \quad F_\lambda(-\frac{\Omega_\lambda}{2}, \ldots) \;, \tag{8.20}$$

with F_λ bounded in the appropriate $H_\varepsilon(..;..)$ (again we have multiplied a factor $\lambda^{-\varepsilon}$ into F_λ). Ω_λ is defined in analogy to (7.9). Obviously, (8.20) remains bounded for $\lambda\to\infty$. The S_N-norms of F_λ also remain bounded. Let

$$G_\lambda(\Omega_\lambda) = \theta(\Omega_\lambda{}^2-4m^2\lambda^{-2}) \ \frac{\sqrt{\Omega_\lambda{}^2-4m^2\lambda^{-2}}}{\Omega_\lambda} \quad .$$

We can estimate, for $\Omega_\lambda{}^2 \geq 4m^2\lambda^{-2}$, $h \geq 0$:

$$\frac{1}{\lambda^\varepsilon\ h^\varepsilon}\sqrt{\Omega_\lambda{}^2-4m^2\lambda^{-2}}\ \left|\frac{1}{\Omega_\lambda+h} - \frac{1}{\Omega_\lambda}\right| = \frac{h^{1-\varepsilon}}{\lambda^\varepsilon}\ \sqrt{\Omega_\lambda{}^2-4m^2\lambda^{-2}}\ \frac{1}{\Omega_\lambda(\Omega_\lambda+h)}$$

$$\leq \frac{h^{1-\varepsilon}}{\lambda^\varepsilon}\ \frac{1}{\Omega_\lambda+h} \leq (2m)^{-\varepsilon}\ \frac{\Omega_\lambda^\varepsilon h^{1-\varepsilon}}{\Omega_\lambda+h} \leq 1 \quad ,$$

and, because of $\sqrt{|x|+|y|}\leq\sqrt{|x|}+\sqrt{|y|}$:

$$\frac{1}{\lambda^\varepsilon\ h^\varepsilon}\ \frac{1}{\Omega_\lambda+h}\ \left|\ \sqrt{(\Omega_\lambda+h)^2-4m^2\lambda^{-2}}\ -\sqrt{\Omega_\lambda{}^2-4m^2\lambda^{-2}}\ \right|$$

$$\leq \frac{1}{\lambda^\varepsilon\ h^\varepsilon}\ \frac{1}{\Omega_\lambda+h}\ \sqrt{h^2+2\Omega_\lambda h} \leq 2^{\frac{1}{2}-\varepsilon}\ m^{-\varepsilon}\ \frac{\Omega_\lambda^\varepsilon h^{\frac{1}{2}-\varepsilon}}{\sqrt{\Omega_\lambda+h}} \leq 2^{\frac{1}{2}-\varepsilon}\ m^{-\varepsilon} \quad .$$

Hence

$$\lambda^{-\varepsilon}\ \frac{G_\lambda(\Omega_\lambda+h) - G_\lambda(\Omega_\lambda)}{h^\varepsilon} \tag{8.21}$$

is bounded in $1\leq\lambda<\infty$, $0\leq h$, $\frac{2m}{\lambda}\leq\Omega_\lambda$: $\lambda^{-\varepsilon}G_\lambda$ is a bounded family in $H_\varepsilon(;\Omega_\lambda)$.

This completes the proof of the theorem in order $\sigma=1$. The constant c_1 can be chosen as $c_1 = 2$, a contribution 1 coming from (8.21) , another 1 from the χ-factors.

We come to the higher orders $\sigma>1$. We assume the theorem to be true in the lower orders $\tau<\sigma$. In analogy to (7.16) we define

$$J^{\lambda}(K,\underset{\sim}{P},P^-_{\lambda},Q)$$

$$= \int dW \prod_1^{\ell} \delta_+(w_i;m/\lambda)\ \tilde{r}^{\lambda}_{\tau}(K_L,\underset{\sim}{P}_L,P^-_{\lambda L},-W)\ \tilde{r}^{\lambda}_{\sigma-\tau}(K_R,\underset{\sim}{P}_R,P^-_{\lambda R},W) \quad . \tag{8.22}$$

Define J^{λ}_1 as in (7.17) , with λ's and ω_{λ}'s inserted in the appropriate places. In particular, a factor $\prod_{R}[\omega_{\lambda}(\underset{\sim}{p}_j)]^{-1}$ must be inserted. By hypothesis of induction $\lambda^{-c_{\tau}\varepsilon} J^{\lambda}_1$ is a bounded family in $H_{\varepsilon_1}(K_R,\underset{\sim}{P}_R,P^-_{\lambda},Q_L,U,W;Q_R,V)$ for $\varepsilon_1<\varepsilon$. Next we form $J^{\lambda}_2(W^-_{\lambda},P^-_{\lambda},Q,U,V)$ by inserting λ's and ω_{λ}'s in suitable places in (7.18). Applying the induction hypothesis again we find that

$$\lambda^{-(c_{\sigma-\tau}+c_{\tau})\varepsilon}\ J^{\lambda}_2$$

is a bounded family in $H_{\varepsilon'}(W^-_{\lambda},P^-_{\lambda},Q,U;V)$ for $\varepsilon'<\varepsilon_1$. The restriction of this product to the mass shell $w^-_{i\lambda} = 0$ exists and remains bounded for $\lambda\to\infty$. We obtain immediately that

$$\lambda^{-(c_{\sigma-\tau}+c_{\tau})\varepsilon}\ J^{\lambda}_3(P^-_{\lambda},Q,U,V)$$

is bounded in $H_{\varepsilon}(P^-_{\lambda},Q,U;V)$, as desired. J^{λ}_3 is defined in analogy to (7.19).

$\tilde{I}^{\lambda}_{\sigma}$ is a sum of terms of the form (8.22), so that we have proved that $\tilde{I}^{\lambda}_{\sigma}$ has the property claimed for $\tilde{r}^{\lambda}_{\sigma}$ in the theorem. If we take the special case in which the sets P , Q, U , V , are empty and note that $\mathcal{S}\subset H_{\varepsilon}$, we see that this property of $\tilde{I}^{\lambda}_{\sigma}$ implies

$$\lim_{\lambda\to\infty} \lambda^{-\varepsilon}\ \tilde{I}^{\lambda}_{\sigma}(k_1,\ldots,k_n) = 0 \quad \text{in } \mathcal{S}' \tag{8.23}$$

for any $\varepsilon>0$. Comparing this with (8.13) written for I_σ instead or r_σ we obtain

$$d(n,\sigma) = SD(I_\sigma) = SD(r_\sigma) \geq - 3n-\sigma(\mu+\nu-4) \quad . \tag{8.24}$$

The equality of $SD(I_\sigma)$ and $SD(r_\sigma)$ has been demonstrated in Chapter V. Using (8.13) again, this time for r_σ , we obtain the desired equation (8.15). We note that this means that in the limit $\frac{m}{\lambda} \to 0$ no infrared singularities of power type occur. Logarithmic singularities are, however, not excluded by this estimate. They occur, indeed, in perturbation theory and are also expected in a rigorous theory from non-perturbative considerations [32].

We must still show that $\tilde{r}_\sigma^\lambda$ satisfies Theorem 7.2 . We use the form of $\tilde{r}_\sigma(P)$ given in (7.23). In analogy to (7.24) we define

$$\tilde{r}^\lambda_{i\sigma\kappa}(P) = \int dQ_0 \; \tilde{I}^\lambda_{i\sigma}(P_0-Q_0,\underset{\sim}{P}) \; \tilde{\eta}_{i\kappa}(Q_0) \quad , \tag{8.25}$$

$$r^\lambda_{i\sigma\kappa}(X) = \eta_{i\kappa}(X^0) \; I^\lambda_{i\sigma}(X) \quad . \tag{8.26}$$

The compact Ξ-support of

$$\Delta^\lambda(\xi_2,..,\xi_n) = \Delta^\lambda(x_1,..,x_n) = r^\lambda_\sigma(X) - \sum_i r^\lambda_{i\sigma\kappa}(X) \tag{8.27}$$

is independent of λ . (8.23) implies $\lim_{\lambda\to\infty} \lambda^{-\varepsilon} r^\lambda_{i\sigma\kappa} = 0$, hence

$$\lim_{\lambda\to\infty} \lambda^{-\varepsilon}\Delta^\lambda(X) = 0 \tag{8.28}$$

for any $\varepsilon>0$. The representation (7.27) becomes

$$\Delta^\lambda(\Xi) = \sum_D D \; F^\lambda_D(\Xi) \quad . \tag{8.29}$$

The F_D^λ are continuous functions with λ-independent compact supports. Moreover, F_D^λ depends continuously on Δ^λ (see the discussion of Theorem 26 in Schwartz [19], p.91), so that

$$\lim_{\lambda\to\infty} \lambda^{-\varepsilon} F_D^\lambda(\Xi) = 0 \qquad (8.30)$$

uniformly in Ξ . The Fourier transform of $\Delta^\lambda(X)$ is

$$\tilde{\Delta}^\lambda(p_1,..,p_n) = \delta^4(\sum_1^n p_i)\, G^\lambda(p_2,..,p_n) \quad . \qquad (8.31)$$

G^λ is the Fourier transform of $\Delta^\lambda(\Xi))$, hence an entire function. From (8.29) and (8.30) we can derive an estimate

$$|G^\lambda(P)| \leq C_\lambda\, \mathcal{P}(P) \quad , \qquad (8.32)$$

where $\mathcal{P}$ is a positive definite polynomial which is independent of λ , and

$$\lim \lambda^{-\varepsilon} C_\lambda = 0$$

for $\varepsilon>0$. Similar estimates hold for the derivatives of G^λ , hence also for its Hölder norms. With the help of this information it can be proved that $\tilde{\Delta}^\lambda$ satisfies Theorem 8.2 . The proof is the same as the one given above for $\tilde{r}_1^\lambda$. The λ-independence of $\tilde{r}_1^\lambda$ was not essential in this proof, boundedness properties of the type (8.32) would have been sufficient. (Of course, the constant c_σ may have to be increased over its value in the simpler case.)

It remains to be shown that $\tilde{r}_{i\sigma\kappa}^\lambda(P)$ satisfies the theorem. A close step-by-step inspection shows that this can be achieved by a faithful repetition of the proof that $\tilde{r}_{i\sigma\kappa}$ satisfies Theorem 7.1 , with the following changes. The functions and distributions $\tilde{I}_{i\sigma}$, f , $\overline{F}$, etc., acquire indices λ , as do the variables P^-, P'^-, etc.

The functions $\chi(p)$, $\chi^{o}(p)$ must be replaced by $\chi(\lambda p)$, $\chi^{o}(\lambda p)$, and factors $[\omega_{\lambda}(p_i)]^{-1}$ must be inserted. Any statement of the form "$f(U,V) \in H_{\varepsilon}(U;V)$" must be replaced by "$\lambda^{-\bar{\varepsilon}} f_{\lambda}(U,V)$ is a bounded family in $H_{\varepsilon}(U;V)$" , with $\bar{\varepsilon}$ a positive number tending to zero together with ε .

It has already been mentioned in Eq. (8.24) that the following is a corollary of Theorem 8.2 .

Theorem 8.3

The scaling degree $d(n,\sigma)$ of $r_{\sigma}(x_1,..,x_n)$ satisfies, for a theory of type (μ,ν) , the inequality

$$d(n,\sigma) \geq -3n - \sigma(\mu+\nu-4) \quad . \tag{8.33}$$

This estimate is of interest only for the non-vanishing $r_{\sigma}(X)$. In any order σ only finitely many r_{σ} are $\neq 0$, and for the others we have $d(n,\sigma) = \infty$.

Note that Theorem 8.3 gives only lower bounds for $d(n,\sigma)$. There are theories for which (8.33) is not optimal. For instance, if $\tilde{r}_1(p_1,..,p_{\mu}) = \delta^4(\Sigma\, p_i) \prod p_i^{\,2}$, then the minimal solution in higher orders is simply $\tilde{r}_{\sigma}(P) = \Pi\, p_i^{\,2}\, \tilde{r}^{o}_{\sigma}(P)$, where r^{o}_{σ} is the solution corresponding to $\tilde{r}^{o}_1(p_1,..,p_{\mu}) = \delta^4(\Sigma\, p_i)$. This gives $d(n,\sigma) = SD(r^{o}_{\sigma}) - 2\mu$ instead of the $d(n,\sigma) = SD(r^{o}_{\sigma}) - 2\mu\sigma$ which would result if (8.33) were an equality. It is also conceivable that a genuine increase of $d(n,\sigma)$ over the value (8.33) might be the result of cancellations between the various terms in I_{σ} . Such cancellations have never been observed, however, and it is unlikely that they really occur.

It has been shown in Chapter V that $r_{\sigma}(x_1,..,x_n)$ is uniquely defined if

$$d(n,\sigma) > -4n+4 \qquad \text{for} \quad n > 2 \quad ,$$
$$d(2,\sigma) \geqslant -6 \quad . \tag{8.34}$$

More generally, the number of ambiguities is closely connected with the difference

$$\delta(n,\sigma) = -4n + 4 - d(n,\sigma) \quad . \tag{8.35}$$

There are no ambiguities if $\delta(n,\sigma) < 0$ for $n > 2$, $\delta(2,\sigma) \leqslant 2$, and the number of ambiguities increases with increasing δ . From (8.33) we obtain

$$\delta(n,\sigma) \leqslant -n + 4 + \sigma(\mu+\nu-4) \quad . \tag{8.36}$$

A theory is called <u>renormalisable</u> if for every $n > 2$ there is a finite $\delta(n)$ such that

$$\delta(n,\sigma) \leqslant \delta(n) \qquad \text{for all } \sigma \geqslant 1 \tag{8.37}$$

and

$$\delta(n) < 0 \tag{8.38}$$

for all n except finitely many. All other theories are called <u>non-renormalisable</u>. In non-renormalisable theories the number of ambiguities increases indefinitely with increasing order σ .

From (8.36) we see that the renormalisability criterion (8.37), (8.38) is satisfied by the theories of type (4,0) and (3,0) . They correspond to the A^4 and A^3 couplings of canonical field theory which we know to be renormalisable in the canonical sense of the word. Barring some unlikely cancellations, these two theories are the only renormalisable ones, apart from trivial variants of the type mentioned in the discussion of Theorem 8.3. Remember that we assume $\mu \geqslant 3$, and note that ν must be even because only polynomials of even degree can be Lorentz invariant.

We devote the rest of the chapter to a short discussion of the two renormalisable theories.

Type (4,0)

We note first that all r_σ with an odd number of arguments vanish:

$$\delta(n,\sigma) = -\infty \quad \text{for } n \text{ odd} \quad . \tag{8.39}$$

The estimate (8.36) becomes for $\mu=4$, $\nu=0$

$$\delta(n,\sigma) \leqslant - n + 4 \quad , \tag{8.40}$$

hence

$$\delta(n) = - n + 4 \quad . \tag{8.41}$$

For $n > 4$ we have $\delta(n) < 0$, hence $r_\sigma(x_1,..,x_n)$, $n > 4$, is uniquely determined in all orders σ by the r_τ of lower order. Because of $\delta(2) = 2$ the same is true for 2-point function.

The case

$$\delta(4,\sigma) = 0 \tag{8.42}$$

is just on the border line where ambiguities start to appear. In every order σ we obtain in $r_\sigma(x_1,..,x_4)$ an ambiguity of the same form

$$c_\sigma \; \delta^4(x_1-x_2) \; \delta^4(x_1-x_3) \; \delta^4(x_1-x_4) \quad , \tag{8.43}$$

c_σ real. This is also the form of $r_1(x_1,..,x_4)$. We can choose

$c_1 = (2\pi)^6$. Any other factor c_1 could be absorbed into the coupling constant g .

The problem of fixing the constants c_σ is closely connected with another basic problem: that of characterising an interaction by criteria that can be formulated outside of perturbation theory. Our procedure of defining a model by specifying its retarded functions in first order is clearly not fully satisfactory. In the present case we can proceed as follows. The 4-point function $\tilde{r}(p_1,..p_4)$ in p-space is of the form (2.27):

$$\tilde{r}(p_1,..,p_4) = \delta^4(p_1+..+p_4)\ \hat{r}(p_1,..,p_4) \quad . \tag{8.44}$$

$\hat{r}$ is defined on the plane $p_1+..+p_4=0$ and is analytic in a neigbourhood of the origin [27] , hence its value at the origin

$$\bar{g} = \hat{r}(0,0,0,0) \tag{8.45}$$

is a meaningful qantity. $\bar{g}$ is real because of the reality of $r(X)$.

We can now define a theory of type (4,0) by demanding that $\bar{g}$ take a prescribed value $\neq 0$ and that all $r(x_1,..,x_n)$ be of the maximal possible s-degree consistent with this value.

Let us apply this criterion in perturbation theory. Let

$$R_\sigma = \hat{r}_\sigma(0,0,0,0) \quad . \tag{8.46}$$

We demand

$$\sum_\sigma R_\sigma\, g^\sigma = \bar{g} \quad , \tag{8.47}$$

with $\bar{g}$ given. Here we are confronted with the problem of defining the "coupling constant" g , i.e. the expansion parameter, which is not determined by the above general characterisation of the theory. The most natural choice clearly consists in taking $\bar{g}$ itself as expansion parameter:

$$g = \bar{g} \quad , \tag{8.48}$$

in which case (8.47) implies

$$R_1 = 1 \ , \quad R_\sigma = 0 \quad \text{for} \quad \sigma > 1 \quad . \tag{8.49}$$

In first order perturbation theory we obtain then exactly the $\tilde{r}_1(p_1,..,p_4)$ from which we started our construction. All the other $\tilde{r}_1(P)$ vanish, since 0 is clearly, among the possible solutions, the one of maximal s-degree. (We apply the maximality requirement separately in every order, disregarding possible cancellations between terms of different order.) The construction of the higher orders proceeds exactly according to our rules, the ambiguities in the 4-point function being fixed by (8.49) .

The choice (8.48) is, however, not the only possible one. It can be generalised to

$$\bar{g} = \bar{g}(g) = \sum_\sigma \bar{g}_\sigma \, g^\sigma \quad , \quad \bar{g}_0 = 0 \quad , \tag{8.50}$$

with $\bar{g}(g)$ a power series in g . The parameters $\bar{g}_\sigma$ can be fixed arbitrarily. (8.49) is then replaced by

$$R_\sigma = \bar{g}_\sigma \quad . \tag{8.51}$$

Let us call the corresponding solution $r'_\sigma(x_1,..,x_n)$. $\tilde{r}'_1(p_1,..,p_4)$ is $= \bar{g}_1 \, \delta^4(\Sigma \, p_i)$, and for the rest we proceed exactly as before, using (8.51) for removing all ambiguity. It is easy to see that the two solutions

$$\sum_\sigma \bar{g}^\sigma \, r_\sigma(X) \qquad (8.52)$$

and

$$\sum_\sigma g^\sigma \, r'_\sigma(X) \qquad (8.53)$$

are equal in the sense of formal power series: if we substitute (8.50) into (8.52) and rearrange the sum into a power series in g, we find that the resulting expansion coefficients $r''_\sigma(X)$ have all the relevant properties of $r'_\sigma(X)$ and are therefore identical with them.

The freedom of the choice of the function $\bar{g}(g)$ in (8.50) is a special case of the freedom that is known in canonical perturbation theory under the name "renormalisation group" (see Ref. [5], Chapter 8). The change from $\bar{g}$ to g corresponds to a finite renormalisation of the coupling constant. A similar freedom exists, in principle, also for the mass m. Instead of treating the mass as a fixed external parameter we could also consider it as a function of the coupling constant: $m = m(g)$. This would, however, lead to a considerable complication of the formalism, due to the explicit occurrence of m (via the Δ_+-function) in the completeness equations (2.36). It is clearly not worthwhile to introduce this complication in our simple model. This is different, however, in models with several particles, where the g-dependence of mass-differences may be physically important. Such a situation is discussed in Ref. [33].

Type (3,0)

This theory corresponds to a A^3-coupling which is known to be meaningless in a rigorous sense [34], because the positivity requirement for the energy is violated. But this difficulty does not manifest itself in perturbation theory. It is therefore possible to look at this theory in our context, and it is even interesting because of some characteristic differences from the (4,0) case which might be of relevance in some more realistic theories.

The estimate (8.36) becomes for $\mu=3$, $\nu=0$

$$\delta(n,\sigma) \leq -n + 4 - \sigma \quad , \qquad (8.54)$$

a sequence that decreases with σ . Except for the obvious case $n=3$, $\sigma=1$, all the $\delta(n,\sigma)$ satisfy the uniqueness criterion (8.34): the theory of type (3,0) is fixed <u>unambiguously</u> by our rules. Such a theory is called "super-renormalisable"*. How does this result agree with the freedom of the renormalisation group that we would expect to be present here as well as in the (4,0) case? We can see what happens if we introduce a new coupling constant g' by

$$g = \sum_{\sigma} g_{\sigma} g'^{\sigma} \quad , \quad g_0 = 0 \quad , \tag{8.55}$$

with fixed coefficients g_σ , and substitute (8.55) into

$$r(X) = \sum_{\sigma} g^{\sigma} r_{\sigma}(X). \tag{8.56}$$

After formal rearrangement we obtain a power series in g' :

$$r(X) = \sum_{\sigma} g'^{\sigma} r'_{\sigma}(X) \quad . \tag{8.57}$$

The r'_σ solve the equations (3.9) including the subsidiary conditions, and r'_1 is of type (3,0) . In the higher orders the condition of maximal s-degree is violated by the addition of terms of the form

$$c_{\sigma} \, \delta^4(x_1-x_2) \, \delta^4(x_1-x_3) \tag{8.58}$$

to the maximally smooth 3-point function. There is no valid reason for discarding such a theory, since the maximality condition in its strict sense of applying separately to each order of perturbation theory is

---

* More generally a theory is called super-renormalisable if only finitely many $\delta(n,\sigma)$ exceed the critical limit 0 where ambiguity sets in.

hardly justifiable. We can consider (8.57) as a perturbative expansion of a theory of type (3,0) in a generalised sense. In this generalised sense we can again characterise the theory outside of perturbation theory.

In analogy to (8.44) we have

$$\hat{r}(p_1,p_2,p_3) = \delta^4(p_1+p_2+p_3)\ \hat{r}(p_1,p_2,p_3) \tag{8.59}$$

with

$$\bar{g} = \hat{r}(0,0,0) \tag{8.60}$$

a finite real number. A theory of type (3,0) is then defined as a theory with a prescribed value of $\bar{g} \neq 0$, such that all $r(X)$ are of the maximal possible s-degree.

In treating this theory we have again the freedom of choosing a coupling constant g with

$$\bar{g} = \Sigma\, g_\sigma\, g^\sigma \tag{8.61}$$

i.e. we can choose the $\bar{g}_\sigma$ freely, apart from the restrictions $\bar{g}_0 = 0$, $\bar{g}_1 \neq 0$ (for simplicity). For any choice of the $\bar{g}_\sigma$ we obtain a perturbation series by our methods with the additional requirement

$$\hat{r}_\sigma(0,0,0) = \bar{g}_\sigma\ . \tag{8.62}$$

The maximality condition is applied within the restricted class of solutions r_σ satisfying (8.62). The r_σ are thereby determined uniquely. The different solutions obtained for different choices of $\bar{g}_\sigma$ agree as formal power series, if the respective coupling constants are expressed correctly in terms of each other, analogously to (8.55). Among the many possibilities will be one corresponding to our more restricted prescription leading to (8.54), namely the one in which we set $\bar{g}_1 = 1$ and use (8.60) as a definition for $\bar{g}_\sigma$, with $\hat{r}_\sigma(p_1,p_2,p_3)$ being the $\hat{r}_\sigma$ of maximal s-degree as constructed in Chapter V . This special choice does not correspond to the natural choice $\bar{g} = g$.

REFERENCES

[1] K. Hepp: Théorie de la renormalisation, Lecture Notes in Physics, Vol. 2 , Springer, Heidelberg 1969 .

[2] J. Valatin: Proc. Roy. Soc. A 225, 535, and A 226, 254 (1954) .

[3] R. A. Brandt: Fortsch. Physik 18, 249 (1970) .

[4] W. Zimmermann, in: Lectures on elementary particles and quantum field theory, Vol. 1 , ed. S. Deser, M. Grisaru, H. Pendleton, MIT Press 1971 .

[5] N. N. Bogoliubov and D. V. Shirkov: Introduction to the theory of quantized fields, Interscience, New York 1959 .

[6] H. Lehmann, K. Symanzik, and W. Zimmermann: Nuovo Cim. 1, 205 (1955).

[7] H. Lehmann, K. Symanzik, and W. Zimmermann: Nuovo Cim. 6, 319 (1957).

[8] V. Glaser, H. Lehmann, and W. Zimmermann: Nuovo Cim. 6, 1122 (1957).

[9] K. Nishijima: Prog. Theor. Phys. 17, 765 (1957); Phys. Rev.119,485 (1960); M. Muraskin and K. Nishijima: Phys. Rev. 122, 331 (1961) .

[10] H. M. Fried: J. Math. Phys. 3, 1107 (1962) .

[11] R. E. Pugh: Ann. Phys. (N.Y) 23, 335 (1963) .

[12] O. Steinmann: Ann. Phys. (N.Y) 29, 76 (1964); 36, 267 (1966) .

[13] R. F. Streater and A. S. Wightman: PCT, spin & statistics, and all that, Benjamin, New York 1964 .

[14] R. Jost: The general theory of qantized fields, Am. Math. Soc., Providence RI 1965 .

[15] K. Hepp: Lectures, Les Houches summer school 1970 (to be published).

[16] H. Epstein and V. Glaser: Informal Meeting on Renormalization Theory, August 1969, Trieste: Report IC/69/121, Int. Centre of Theoretical Physics, Trieste.

[17] H. Epstein and V. Glaser: Prépublications de la R.C.P. no. 25, vol. 11 , Institut de mathématique, Université de Strasbourg 1970; also: CERN preprint TH 1156 (1970) .

[18] O. Steinmann: Comm. Math. Phys. 10, 245 (1968)

[19] L. Schwartz: Théorie des distributions, Hermann, Paris 1966 .

[20] K. Hepp: Comm. Math. Phys. 1, 95 (1965) .

[21] W. Schneider: Helv. Phys. Acta 39, 81 (1966)

[22] N. I. Muskhelishvili: Singular integral equations, P. Noordhoff N.V., Groningen 1953 .

[23] L. D. Faddeev: Mathematical aspects of the three-body problem in the quantum scattering theory, Israel program for scientific translations, Jerusalem 1965 .

[24] O. Steinmann: Helv. Phys. Acta 33, 257 and 347 (1960) .

[25] H. Araki: J. Math. Phys. 2, 163 (1961) .

[26] D. Ruelle: Nuovo Cim. 19, 356 (1961) .

[27] H. Epstein, in: Axiomatic field theory (Brandeis Summer School 1965), Gordon and Breach, New York 1966 .

[28] P. D. Methée: Comment. Math. Helv. 28, 225 (1954)

[29] L. Garding: Nuovo Cim. Suppl. 14, 45 (1959) .

[30] A. Lichnerowicz: Algèbre et analyse linéaires, p. 123, Masson & Cie., Paris 1947 .

[31] O. Steinmann: Comm. Math. Phys. 18, 179 (1970) .

[32] K. Symanzik: Comm. Math. Phys. 18, 227 (1970) .

[33] O. Steinmann: Comm. Math. Phys. 15, 133 (1969) .

[34] K. Osterwalder: Fortsch. Physik (to appear) .

Lecture Notes in Physics

Bisher erschienen / Already published

Vol. 1: J. C. Erdmann, Wärmeleitung in Kristallen, theoretische Grundlagen und fortgeschrittene experimentelle Methoden. 1969. DM 20,–

Vol. 2: K. Hepp, Théorie de la renormalisation. 1969. DM 18,–

Vol. 3: A. Martin, Scattering Theory: Unitarity, Analyticity and Crossing. 1969. DM 14,–

Vol. 4: G. Ludwig, Deutung des Begriffs physikalische Theorie und axiomatische Grundlegung der Hilbertraumstruktur der Quantenmechanik durch Hauptsätze des Messens. 1970. DM 28,–

Vol. 5: M. Schaaf, The Reduction of the Product of Two Irreducible Unitary Representations of the Proper Orthochronous Quantummechanical Poincaré Group. 1970. DM 14,–

Vol. 6: Group Representations in Mathematics and Physics. Edited by V. Bargmann. 1970. DM 24,–

Vol. 7: R. Balescu, J. L. Lebowitz, I. Prigogine, P. Résibois, Z. W. Salsburg, Lectures in Statistical Physics. 1971. DM 18,–

Vol. 8: Proceedings of the Second International Conference on Numerical Methods in Fluid Dynamics. Edited by M. Holt. 1971. DM 28,–

Vol. 9: D. W. Robinson, The Thermodynamic Pressure in Quantum Statistical Mechanics. 1971. DM 14,–

Vol. 10: J. M. Stewart, Non-Equilibrium Relativistic Kinetic Theory. 1971. DM 14,–

Vol. 11: O. Steinmann, Perturbation Expansions in Axiomatic Field Theory. 1971. DM 14,–

Selected Issues from

Lecture Notes in Mathematics

Vol. 7: Ph. Tondeur, Introduction to Lie Groups and Transformation Groups. Second edition. VIII, 176 pages. 1969. DM 14,–

Vol. 40: J. Tits, Tabellen zu den einfachen Lie Gruppen und ihren Darstellungen. VI, 53 Seiten. 1967. DM 6,80

Vol. 52: D. J. Simms, Lie Groups and Quantum Mechanics. IV, 90 pages. 1968. DM 8,–

Vol. 55: D. Gromoll, W. Klingenberg und W. Meyer, Riemannsche Geometrie im Großen. VI, 287 Seiten. 1968. DM 20,–

Vol. 56: K. Floret und J. Wloka, Einführung in die Theorie der lokalkonvexen Räume. VIII, 194 Seiten. 1968. DM 16,–

Vol. 75: G. Lumer, Algèbres de fonctions et espaces de Hardy. VI, 80 pages. 1968. DM 8,–

Vol. 81: J.-P. Eckmann et M. Guenin, Méthodes Algébriques en Mécanique Statistique. VI, 131 pages. 1969. DM 12,–

Vol. 82: J. Wloka, Grundräume und verallgemeinerte Funktionen. VIII, 131 Seiten. 1969. DM 12,–

Vol. 91: N. N. Janenko, Die Zwischenschrittmethode zur Lösung mehrdimensionaler Probleme der mathematischen Physik. VIII, 194 Seiten. 1969. DM 16,80

Vol. 89: Probability and Information Theory. Edited by M. Behara, K. Krickeberg and J. Wolfowitz. IV, 256 pages. 1969. DM 18,–

Vol. 103: Lectures in Modern Analysis and Applications I. Edited by C. T. Taam. VII, 162 pages. 1969. DM 12,–

Vol. 128: M. Takesaki, Tomita's Theory of Modular Hilbert Algebras and its Applications. II, 123 pages. 1970. DM 10,–

Vol. 140: Lectures in Modern Analysis and Applications II. Edited by C. T. Taam. VI, 119, pages. 1970. DM 10,–

Vol. 144: Seminar on Differential Equations and Dynamical Systems II. Edited by J. A. Yorke. VIII, 268 pages. 1970. DM 20,–

Vol. 167: Lavrentiev, Romanov and Vasiliev, Multidimensional Inverse Problems for Differential Equations. V, 59 pages. 1970. DM 10,–

Vol. 170: Lectures in Modern Analysis and Applications III. Edited by C. T. Taam. VI, 213 pages. 1970. DM 18,–

Vol. 183: Analytic Theory Differential Equations. Edited by P. F. Hsieh and A. W. J. Stoddart. VI, 225 pages. 1971. DM 20,–

Vol. 190: Martingales. A Report on a Meeting at Oberwolfach, May 17-23, 1970. Edited by H. Dinges. V, 75 pages. 1971. DM 12,–

Vol. 192: Proceedings of Liverpool Singularities – Symposium I. Edited by C. T. C. Wall. V, 319 pages. 1971. DM 24,–

Vol. 193: Symposium on the Theory of Numerical Analysis. Edited by J. Ll. Morris. VI, 152 pages. 1971. DM 16,–

Vol. 197: Manifolds – Amsterdam 1970. Edited by N. H. Kuiper. V, 231 pages. 1971. DM 20,–

Vol. 198: M. Hervé, Analytic and Plurisubharmonic Functions in Finite and Infinite Dimensional Spaces. VI, 90 pages. 1971. DM 16,–

Vol. 206: Symposium on Differential Equations and Dynamical Systems. Edited by D. Chillingworth. XI, 173 pages. 1971. DM 16,–

Beschaffenheit der Manuskripte

Die Manuskripte werden photomechanisch vervielfältigt; sie müssen daher in sauberer Schreibmaschinenschrift mit ausreichend großer Type geschrieben sein. Handschriftliche Formeln bitte nur mit schwarzer Tusche eintragen. Notwendige Korrekturen sind bei dem bereits geschriebenen Text entweder durch Überkleben des alten Textes vorzunehmen oder aber müssen die zu korrigierenden Stellen mit weißem Korrekturlack abgedeckt werden. Die reproduktionsfähigen Abbildungen (in Originalgröße) sollen in den Text eingeklebt werden. Falls das Manuskript oder Teile desselben neu geschrieben werden müssen, ist der Verlag bereit, dem Autor bei Erscheinen seines Bandes einen angemessenen Betrag zu zahlen. Die Autoren erhalten 50 Freiexemplare.

Zur Erreichung eines möglichst optimalen Reproduktionsergebnisses ist es erwünscht, daß bei der vorgesehenen Verkleinerung der Manuskripte der Text auf einer Seite in der Breite möglichst 18 cm und in der Höhe 26,5 cm nicht überschreitet. Entsprechende Satzspiegelvordrucke werden vom Verlag gern auf Anforderung zur Verfügung gestellt.

Manuskripte, in englischer, deutscher oder französischer Sprache abgefaßt, sind einzureichen bei: Springer-Verlag, 6900 Heidelberg, Postfach 1780.

Cette série a pour but de donner des informations rapides, de niveau élevé, sur des développements récents en physique, aussi bien dans la recherche que dans l'enseignement supérieur. On prévoit de publier.

1. des versions préliminaires de travaux originaux et de monographies
2. des cours spéciaux portant sur un domaine nouveau ou sur des aspects nouveaux de domaines classiques
3. des rapports de séminaires
4. des conférences faites lors de congrès ou de colloques

En outre il est prévu de publier dans cette série, si la demande le justifie, des rapports de séminaires et des cours multicopiés ailleurs mais déjà épuisés.

Dans l'intérêt d'une diffusion rapide, les contributions auront souvent un caractère provisoire; le cas échéant, les démonstrations ne seront données que dans les grandes lignes. Les travaux présentés pourront également paraître ailleurs. Une réserve suffisante d'exemplaires sera toujours disponible. En permettant aux personnes intéressées d'être informées plus rapidement, les éditeurs Springer espèrent, par cette série de «prépublications», rendre d'appréciables services aux instituts de physique. Les annonces dans les revues spécialisées, les inscriptions aux catalogues et les copyrights rendront plus facile aux bibliothèques la tâche de réunir une documentation complète.

Présentation des manuscrits

Les manuscrits, étant reproduits par procédé photomécanique, doivent être soigneusement dactylographiés type assez grand. Il est recommandé d'écrire à l'encre de Chine noire les formules non dactylographiées. Les corrections nécessaires doivent être effectuées soit par collage du nouveau texte sur l'ancien soit en recouvrant les endroits à corriger par du vernis correcteur blanc. Les illustrations; en dimension originale, préparées pour reproduction sont à insérer dans le texte. S'il s'avère nécessaire d'écrire de nouveau le manuscrit, soit complètement, soit en partie, la maison d'édition se déclare prête à verser à l'auteur, lors de la parution du volume, le montant des frais correspondants. Les auteurs recoivent 50 exemplaires gratuits.

Pour obtenir une reproduction optimale il est désirable que le texte dactylographié sur une page ne dépasse pas 26,5 cm en hauteur et 18 cm en largeur. Sur demande la maison d'edition met à la disposition des auteurs du papier spécialement preparé.

Les manuscrits en anglais, allemand ou français peuvent être adressés à Springer-Verlag, 6900 Heidelberg, Postfach 1780.